Dance, Space and Subjectivity

Dance, Space and Subjectivity

Valerie A. Briginshaw
Formerly Professor of Dance Studies
University of Chichester

© Valerie A. Briginshaw 2001, 2009

All rights reserved. No reproduction, copy or transmission of this publication may be made without written permission.

No portion of this publication may be reproduced, copied or transmitted save with written permission or in accordance with the provisions of the Copyright, Designs and Patents Act 1988, or under the terms of any licence permitting limited copying issued by the Copyright Licensing Agency, Saffron House, 6–10 Kirby Street, London EC1N 8TS.

Any person who does any unauthorized act in relation to this publication may be liable to criminal prosecution and civil claims for damages.

The author has asserted her right to be identified as the author of this work in accordance with the Copyright, Designs and Patents Act 1988.

First published in hardback 2001
First published in paperback 2009 by
PALGRAVE MACMILLAN

Palgrave Macmillan in the UK is an imprint of Macmillan Publishers Limited, registered in England, company number 785998, of Houndmills, Basingstoke, Hampshire RG21 6XS.

Palgrave Macmillan in the US is a division of St Martin's Press LLC, 175 Fifth Avenue, New York, NY 10010.

Palgrave Macmillan is the global academic imprint of the above companies and has companies and representatives throughout the world.

Palgrave® and Macmillan® are registered trademarks in the United States, the United Kingdom, Europe and other countries.

ISBN-13: 978-0-333-91973-6 hardback
ISBN-13: 978-0-230-22979-2 paperback

This book is printed on paper suitable for recycling and made from fully managed and sustained forest sources. Logging, pulping and manufacturing processes are expected to conform to the environmental regulations of the country of origin.

A catalogue record for this book is available from the British Library.

Library of Congress Cataloging-in-Publication Data
Briginshaw, Valerie A.
   Dance, space, and subjectivity / Valerie Briginshaw.
     p. cm.
   Includes bibliographical references and index.
   ISBN 978-0-333-91973-6 (cloth)   978-0-230-22979-2 (pbk)
     1. Dance—Reviews. 2. Postmodernism. I. Title.
   GV1594.B75 2001
   792.8—dc21
                                                      2001021214

10 9 8 7 6 5 4 3 2 1
18 17 16 15 14 13 12 11 10 09

Printed and bound in Great Britain by
CPI Antony Rowe, Chippenham and Eastbourne

to the memory of my parents
Vera Briginshaw (née Higham)
and
George Maitland Briginshaw

Contents

List of Illustrations	xi
Preface to Second Edition	xiii
Preface	xv
Acknowledgements	xviii

1 Introduction	**1**
Where bodies meet space	2
Currency of ideas about space	5
Questions of subjectivity	6
Reading dances	7
Bounded bodies	9
In-between spaces	14
Inside/outside interfaces	17

PART I   Constructions of Space and Subjectivity

2 Travel metaphors in dance – gendered constructions of travel, spaces and subjects	**27**
Introduction	27
Metaphors of travel in postmodern discourse	28
Travel and gender	29
Space, power and gender	30
Suggestions of travel in the dances	32
Gendering of travel, space and subjects in the dance films	35
Reappropriating travel metaphors for new subjectivities	40
Conclusion	42
3 Transforming city spaces and subjects	**43**
Introduction	43
Sixties precedents	44
Muurwerk and *Step in Time Girls*	49
Constructed spaces	50
Mutual definition of bodies and cities	51
Conclusion	56

4 Coastal constructions in Lea Anderson's *Out on the Windy Beach* — 59

Introduction — 59
Eroticized bodies and spaces — 63
Bodies and boundaries — 65
Seaside surrealism — 68
Conclusion — 73

PART II Dancing in the 'In-Between Spaces'

5 Desire spatialized differently in dances that can be read as lesbian — 77

Introduction — 77
Lesbian desire refigured — 78
The performance of gender and sexuality — 80
The dances – *Reservaat* (1988), *Between/Outside* (1999), *Virginia Minx at Play* (1993) and *Homeward Bound* (1997) — 81
Surfaces in contact — 83
Becoming/transformations in *Reservaat* and *Homeward Bound* — 85
Machinic assemblages in *Between/Outside* and *Reservaat* — 88
Polymorphous perversity and multiplicities in *Virginia Minx at Play* — 92
In-between spaces – spatial locations and private/public boundaries — 94
Conclusion — 95

6 Hybridity and nomadic subjectivity in Shobana Jeyasingh's *Duets with Automobiles* — 97

Introduction — 97
Hybrid spaces between East and West — 99
Female solidarity and nomadic subjectivity — 102
Dancers and buildings and the spaces in between — 104
Mutual construction of bodies and spaces — 106
Nomadic subjects in cities — 108
Conclusion — 109

7 Crossing the (black) Atlantic: spatial and temporal displacements in Meredith Monk's *Ellis Island* and Jonzi D's *Aeroplane Man* — 112

Introduction — 112
Meredith Monk and Jonzi D — 114

The constructed nature of identity	117
Technologies of power – subjectification, normalization and examination	123
The role of language in operations of power	130
Spatial containment, borders and territorialization	132
Conclusion	134

PART III Inside/Outside Bodies and Spaces

8 Fleshy corporealities in Trisha Brown's *If You Couldn't See Me*, Lea Anderson's *Joan* and Yolande Snaith's *Blind Faith* — 139

Introduction	139
The theories of Deleuze, Bordo and Kristeva	141
The dances	144
Flesh	146
Fluids	153
Folds	155
Conclusion	161

9 'Carnivalesque' subversions in Mark Morris' *Dogtown*, Liz Aggiss' *Grotesque Dancer* and Emilyn Claid's *Across Your Heart* — 162

Introduction	162
The dances	164
Grotesque challenges to boundaries	166
Excessive overflows	171
Carnivalesque parody	175
Grotesque and carnivalesque interactivity with the world	178
Conclusion	181

10 Architectural spaces in the choreography of William Forsythe and De Keersmaeker's *Rosas Danst Rosas* — 183

Introduction	183
The choreographers and dances	186
Disrupting the single viewpoint of perspective	189
Challenging notions of a separated self	192
The role of the visual in the construction of bounded and gendered subjectivity	195
Vision and reason – the dominance and construction of the eye/'I'	198
The ideological nature of visualization	200
Conclusion	204

Appendix	207
Notes	209
Bibliography	213
Index	223

List of Illustrations

Cover illustration Film still of Martine Berghuijs and Pépé Smit in *Reservaat* courtesy of Cinenova Film & Video Distribution.
Plate 1 Film still of Roxanne Huilmand in *Muurwerk* courtesy of argos international film distributors.
Plate 2 Photograph of Yolande Snaith in *Step in Time Girls* by Ross MacGibbon reproduced with his permission.
Plate 3 Photograph of *Out on the Windy Beach* by the author.
Plate 4 Photograph of Emilyn Claid in the 'Watch Me Witch You' section from *Virginia Minx at Play* by Eleni Leoussi reproduced with her permission.
Plate 5 Photograph of Sarah Spanton in *Homeward Bound* by Michelle Atherton reproduced with her permission.
Plate 6 Still from *Duets with Automobiles* from the series 'Dance for the Camera' courtesy of Arts Council/BBC.
Plate 7 Photograph of *Ellis Island* by Bob Rosen reproduced with his permission.
Plate 8 Photograph of Trisha Brown in *If You Couldn't See Me* by Joanne Savio reproduced with her permission.
Plate 9 Photograph of *The Anatomy Lesson of Dr Tulp* (1632) by Rembrandt courtesy of Mauritshuis, The Hague.
Plate 10 Still from *Joan* courtesy of MJW Productions.
Plate 11 Photograph of Andreas Vesalius's illustration The Fifth Plate of the Muscles from the Second Book of the *Fabrica* reproduced with permission of Dover Publications Inc.
Plate 12 Still from *Blind Faith* courtesy of Yolande Snaith Theatredance.
Plate 13 Photograph of Liz Aggiss in *Grotesque Dancer* reproduced with permission of Divas.
Plate 14 Photograph of CandoCo Dance Company in *Across Your Heart* by Chris Nash reproduced with his permission.
Plate 15 Photograph of *Rosas Danst Rosas* by Herman Sorgeloos reproduced with his permission.
Plate 16 Photograph of *Enemy in the Figure* by Dominik Mentzos reproduced with his permission.

Preface to Second Edition

Despite being published eight years ago in 2001, the core ideas in *Dance, Space and Subjectivity* concerned with the ways in which dances construct space and subjectivity and the meanings that result, are still pertinent today. No other texts in dance studies have dealt since, at length, with ideas about the spatiality of subjectivity in dance in any detail. The premises and arguments on which the original text was based are still current and relevant as is the content in the form of the ideas explored and the dances discussed. Most of the choreographers whose dance works are examined are important national and international figures whose choreography is still very much at the centre of current debate in dance scholarship. They continue to occupy an important position in the dance world today as they did when the first edition appeared, and they remain worthy of serious study. The close readings of dances in *Dance, Space and Subjectivity* can also serve as useful models for exploring dance in-depth in practice and in writing within dance and performance studies courses.

The revisions I have made for this edition are therefore relatively minor and largely in the form of correcting one or two minor typographical and grammatical errors that slipped into the first edition, and updating information in the text, particularly in Appendix 1. This has been completely re-written since it indicates details of the availability of recordings of dance works discussed.

I thought long and hard about the use of the term 'postmodern' in this second edition. It is a contentious term that was the subject of much debate throughout the 1980s and 90s in which I and others participated (see Briginshaw, 1988, 1996). Re-reading my contributions to the debates and the very first footnote of this text, where I qualify my use of the term with reference to dance, with which I still concur, for me, these clarify understanding of the term in general and my use of it in particular. It was through my engagement with various facets of what was termed 'postmodern theory' that I became aware of ways in which the notion of 'subjectivity' could be opened up. Discussions of the 'crisis in subjectivity', often deemed 'postmodern' at the time, heralded further discussions and debates about what subjectivity is, which have informed explorations that are at the heart of this text. It is my contention that readers will inevitably come across the term 'postmodern' in relation to the dances and theories discussed here in other literature, and that my qualified use of it here contextualizes and clarifies the issues and debates and aids understanding. I have therefore decided to let it stand.

Revisions I have made have resulted from changes in my thinking over the past eight years consistent with my statement in the original preface that 'my subjectivity and that of any writer or dance critic is made up of several layers that are continually changing over time and in different spaces' (p. xvi). As a result I have changed all references to the notion of theories being 'applied to' dance to less deterministic and closed language in order to emphasize the importantly open nature of the research with which I am engaged. Making use of terms such as 'exploration' and 'discussion' of theories 'alongside dance' in order to 'inform debates', I feel, is a more accurate and helpful way of describing the work undertaken in the text. The value and importance of openness in dance and in writing, Ramsay Burt and I emphasize and explore in more detail in our recently published *Writing Dancing Together*. The timing of this second edition of *Dance, Space and Subjectivity* has also allowed me to refer readers to the former text where ideas such as 'mutation' and 'unfaithful repetition', discussed here in Chapters 4 and 5 respectively, are explored in more detail, with reference, for example, to Lea Anderson's work.

The publishers' request for me to consider a second edition of *Dance, Space and Subjectivity* coincided with my decision to retire from the University of Chichester and most of my academic engagements, research and writing in order to pursue other activities. These include voluntary work in developing countries which is in some ways a continuation of the social and political commitment that informed the writing of this book. This commitment, outlined in the original preface, is the *raison d'etre* behind the text and behind this paperback edition, which I hope will render it more accessible for purchase to a broader readership.

Chichester, 2009

Preface

In drawing together the issues of dance, space and subjectivity, this book centres on dance analysis, but it also draws on a number of different fields which extend beyond dance studies. For the past two decades I have been engaged in research that has taken as its focus the field of dance analysis, evidenced in my contributions to publications in the eighties and nineties, including the book *Dance Analysis: Theory and Practice*. Since then I have become increasingly interested in contemporary critical theories, beginning with feminist theory, and more recently with postmodern, postcolonial, queer and post-structuralist theories. I find these theories compelling because of the ways in which they bring to the surface issues and debates about the nature of human existence and human relationships, that have remained hidden for far too long.

These issues and debates concern similarities and differences between people that begin with where, when and to whom we are born. They are fundamental because they are political; they result in power differentials and discrimination. These similarities and differences evolve in complexity for each of us with passages through different times and spaces. Crucially these similarities and differences, which make up our subjectivity, determine our life experiences. All of this may seem very obvious, but when we look around and see large numbers of people whose life experiences and expectations are so much less than those of others, somehow it doesn't seem fair. I believe, perhaps somewhat naively, that this lack of justice or fairness can be alleviated with the greater understanding of its root causes and factors that perpetuate it, that feminist, postmodern, postcolonial, queer and post-structuralist theories can provide.

My knowledge and understanding of these theories has been continually challenged and stretched by my own and others' attempts to explore them alongside dance and to explore dance alongside the theories. This book sits happily in the middle of this open field of mutual application and engagement where insights gained in one area of cultural practice or theory can illuminate another. These mutual illuminations, which open up rather than close down issues and areas of debate and contention, for me, are extremely productive. The major imperative behind this book is to share this productivity.

Over the years my focus in dance analysis has been on those postmodern dances which lend themselves to analysis informed by contemporary critical theories of this kind. This book consists of close readings of selected examples of postmodern dance that I have chosen because of the ways in which the dancing bodies in each construct subjectivity and the different roles that space plays. I therefore, develop frameworks for analysis in the book which relate to other frameworks focusing on the contributions of space and the dancing bodies to constructions of subjectivity. One main aim in all this is to call into question traditional notions of subjectivity which see it as fixed, and to suggest more interesting alternatives which recognize its capacities for change. In this sense it is important to stress the complex and changing nature of my own subjectivity which plays an important role in the writing. My subjectivity and that of any writer or dance critic is made up of several layers that are continually changing over time and in different spaces. This is why, although I present particular readings and interpretations of the dances I write about, I stress that these are not the only possible readings. I hope that these will open up the dances and provoke and stimulate further interpretations of these dances and maybe even of works with similar qualities in other cultural forms.

Another incentive for illustrating the reciprocity of dance and a range of different theories in this book is to demonstrate the broad relevance of dance as a subject of study that can illuminate other areas of cultural practice. In this sense this book should be of interest to a readership that extends beyond students and researchers of dance. Many of the dances that feature, for example, are films or dance videos and so the analysis draws on film theory. In its readings of postmodern dances the book engages with topics of cultural study as diverse as travel, the city, beaches as leisure spaces, environmental change, the spatialization of desire, displacement and cultural identity, medieval and Renaissance art and literature and deconstructive architecture. In this sense it should appeal to students and scholars in the fields of cultural history and geography, visual culture, feminism, postcolonial, queer and gender studies as well as cultural and critical theory more generally. At the heart of the analyses of all the dances, however, is a focus on bodies and space. The treatment of the ways in which performing bodies engage and interact with actual spaces and embody ideas associated with metaphorical spaces, and how this contributes to constructions of subjectivity, should also be of interest to students and researchers within the fields of performance, theatre and media studies. Finally, it is my hope that in demonstrating the attraction, relevance and diversity of the dances through my analyses, the profile of dance as a cultural practice worthy of serious attention for cultural studies in general, will be raised.

Acknowledgements

I would like to acknowledge and thank the following people for their invaluable help, advice, guidance and support in the writing of this book. Special mention goes to Ramsay Burt who over many years as a friend and colleague has shared ideas, challenged and stretched my thinking, encouraged, prompted and supported me. He read most of the chapters at various stages in their development, and he read the manuscript in its entirety. His constructive criticism has been invaluable. Thank you Ramsay. My good friend and colleague Geoff Seale also read the whole manuscript. As he comes from outside the field of Dance his detailed comments and suggestions tended to touch on a broad range of issues and to stimulate revisions which I hope have made the book more user friendly, particularly for those readers not necessarily acquainted with the dances or theories discussed. Other friends and colleagues have read drafts or early versions of chapters and given feedback. They are Christy Adair (Chapters 5 and 7), Johannes Birringer (Chapter 1), Ruth Chandler (Chapters 5 and 8), Emilyn Claid (Chapters 5 and 9), Diana Crampton (Chapter 8), Isla Duncan (Chapters 2 and 5), Ben Noys (Chapter 9), Ann Nugent (Chapter 10), and Sarah Rubidge (Chapter 6). I should also like to thank Daniela Adriana Pegorer, for advice concerning the tango for Chapter 5 and Saul Keyworth for reading Chapter 7 and giving advice concerning break dance and hip hop. I am grateful to all of the above for their advice but I take full responsibility for what remains.

Some of the chapters or parts thereof have been presented as conference papers, research seminar papers, lectures or have been published elsewhere. Parts of Chapter 2 were presented at the *Border Tensions* Conference at the University of Surrey in 1995. Chapter 3 grew out of a conference paper presented at the *City Limits* Conference at Staffordshire University in 1996, a chapter Helen Thomas asked me to write for her edited collection *Dance in the City* published by Macmillan (– now Palgrave) in 1997, and a chapter in Bell and Haddour's edited collection *City Visions* published by Pearson Education Limited in 2000. Chapter 5 grew out of two conference papers: the first presented at a panel on 'Dance and Sexuality' convened by Susan Manning for the Society of Dance History Scholars Conference at the University of Oregon in 1998, and the second presented at the Society of Dance History Scholars Conference at the University of New Mexico in 1999. I was invited to give an early version of this conference paper at Surrey University School of Performing Arts Research Seminar Week in 1999, and to give a version of Chapter 5 as a lecture to MA students on the 'Gender and Sexuality in Performance' module at the Tisch School of Performing Arts, New York University. I should like to thank both institutions for their invitations, and their respective students and staff for their useful feedback.

Chapter 6 grew out of a paper given at the *Confluences* conference at the University of Cape Town in 1997, and the chapter in *City Visions* published by Pearson Education Limited in 2000 mentioned above. Chapter 8 grew out of an article written for the journal *Visual Culture in Britain* published in 2000, and part of this material was presented at the conference *Visual Culture in a Changing Society: Britain 1940–2000* organized by the journal at the University of Northumbria in 2000. I should like to thank the journal editor, Ysanne Holt, for permission to publish, and her selected readers for their feedback. I also acknowledge permission to publish material in Chapters 3 and 6 from Pearson Education Limited. Chapter 10 grew out of two lectures given to MA Performing Arts students at De Montfort University, Leicester in 1999 and to an audience of architecture and design students and staff of the Hochschule Anhalt (FH) at the Bauhaus in Dessau, in 1999. I should like to thank Ramsay Burt and Professor Lisa Stybor respectively for inviting me to give the lectures, as well as the staff and students who attended and gave me valuable feedback.

Some of the material in the book has also been presented as part of University College Chichester Research Seminar Programmes for the Dance and Women's Studies Departments and the Postmodern Studies Research Centre. Thanks go to all who attended and gave their comments. I should like to acknowledge support from colleagues and students at University College Chichester, particularly in the Dance Department and the Postmodern Studies Research Centre Critical Theory Reading Group and Ruth Twiss, the Dance Subject Librarian at University College Chichester for help with searches and various other obscure requests. Thanks also go to Paul Burt and Damian Wiles, technicians in the Media Centre, University College Chichester, for help with reproduction of stills from videos for Plates 10 and 12. Financial support for my attendance at conferences and lectures overseas has been generously provided by University College Chichester's Dance Department research budget.

I should like to thank copyright holders for their permission to reproduce the visual images in the book. Detailed acknowledgements are made with the captions.

Thanks go to all choreographers, filmmakers and dancers whose work is discussed in the book. Their stimulating work has inspired my writing. To those choreographers who have read and/or discussed my writing at various stages, namely Liz Aggiss, Lea Anderson, Emilyn Claid, Shobana Jeyasingh, Lucille Power, Yolande Snaith and Sarah Spanton; thanks go to them for their valuable time and comments.

This manuscript was submitted for a Ph.D. by publication to the University of Southampton. Its examination culminated in a successful *viva voce* in November 2000. My three examiners Dr Cora Kaplan of the University of Southampton, Dr Lizbeth Goodman of the University of Surrey and Professor Stephanie Jordan of Roehampton Institute, provided valuable advice regarding revisions that would strengthen the text in its published form, which I would like to acknowledge.

Lastly, I should like to thank all those close friends who have in various ways lived with me and my idiosyncracies through the process of putting this book together. Particular mentions go to Charmian Gradwell for stimulating and challenging discussions of some of the content, and for cooking me delicious meals to sustain my writing while on a writing retreat holiday, and to Ginny Levett for reminding me, while on walks along beaches, that shorelines are special kinds of borders and beaches are special kinds of liminoid spaces.

1
Introduction

This book is about relations between bodies and space in dance and the role they play in constructing subjectivity. Why, at the beginning of the new millenium, are relationships between dance, space and subjectivity so important? The postmodern debates about 'crises of subjectivity', that were current at the turn of the twentieth century, raised questions about who 'we' are, and our relation to the world we live in. Although voiced increasingly sceptically now, these questions are still, it seems to me, just as pressing as ever. Postmodern dance contributes to these debates vitally and imaginatively, because it is constantly engaging and negotiating with body/space relations in immediate and challenging ways.[1] I use the oblique (/) here between 'body' and 'space' to indicate the conjunction of two concepts creating an interface. It allows the possibility of rethinking concepts or ideas normally seen as separated, as interconnected. Thinking things differently in this way is a key strategy employed throughout the book, often indicated by the use of the slash.

The conjunction of bodies and spaces is important because it is through this interface, through our material bodies being in contact with space, that we perceive the world around us and relations to that world. Exploration of these relations and debates is particularly pertinent right now because of the openness and scope that postmodern approaches have brought to questions of existence, identity and subjectivity. The limitations of the perspective of an ideal, unified, male, Western, white subject have been exposed – presenting the point of view from this subject position as 'the way things are' is no longer acceptable. Many other possibilities are now viable. Through in-depth analyses of selected postmodern dances, informed by developments and debates in current critical theory, this book identifies some of these other possibilities of subjectivity. In dance the limits and extent of subjectivity are actually and metaphorically exposed when and where bodies meet space. This is perhaps why there has been an explosion of interest in bodies and space and bodily and spatial metaphors in current cultural theory. The debates and issues that have emerged have not to date been examined in dance in any detail, yet dance is an obvious medium for the exploration and imagination of what is going on at these limits.

Where bodies meet space

An example of the significance of body/space relations in dance can be seen in the opening moments of *Outside/In* (1995), a dance film commissioned by the Arts Council and BBC2, made by Margaret Williams and choreographed by Victoria Marks, for CandoCo Dance Company. The camera pans slowly along a line of heads and shoulders of dancers from left to right as they pass small intimate gestures, such as blown and actual kisses, from one to the next. Each dancer contacts the next in her or his own way showing individuality, but the touch in each case is a shared gesture of intimacy and affection. The appearance of the dancers, who they touch, how they make contact and where they are, carry a whole gamut of connotations, associations and ambiguities suggesting different possible readings. These are inherent in the various relations of the dancers' bodies and the kinds of bodies they are, to the space around, and importantly, between them.

The phrase begins with a young white woman raising her palm to touch her chin and blowing air into the face of an older white woman to her left across her fingertips which touch the second woman's chin. She reels back on impact smiling with pleasure while continuing to look at the younger woman. She then slowly turns to look at a young dark man to her left and blows air in his ear. His head jerks on receipt of the 'gift' and he puts his finger in his other ear as though clearing it to let the air pass through. He then turns to face camera, puckers his lips ready for a kiss and after rolling his eyes to right and left, he places a kiss on the lips of a blond young man to his left, also with puckered lips. Turning to face camera, the blond man licks his lips with pleasure, looks to his left slowly, and touches foreheads with a young white woman next to him, they both look down into her cupped hands into which he blows. She sniffs the air and lifts up her head, inhaling, with her eyes closed momentarily, as if to savour the experience. She then lowers her head, and as the head and shoulders of an older white man are laid on her lap, she blows into his open mouth.

Closing his mouth, he raises his head, helped and supported by her hand until he is facing the young white woman to his left who was first seen at the beginning of the line. He rubs noses with her three times turning her face to camera as he does so, and another series of intimate gestures begins.

The spaces between and around the dancers set up particular resonances. Throughout the sequence the performers come very close to each other, often touching and maintaining eye contact, suggesting intimacy and affection, which is enhanced by the slow, soft accompanying music and camera close-ups. The head and shoulder shots focus attention on the kinds of people that are touching; their faces foreground marks of identity such as age, gender and 'race'. Intimate gestures such as kisses and nose rubbing normally occur between youngish people of different genders and the same 'race'. This is not the case here, contact is made between young and old, people of different 'racial' origins, and not only between male and female, but also between females and males, possibly suggesting different sexualities. The norm is challenged; other possibilities of who is allowed to kiss or intimately touch who, of what is 'acceptable', are presented. Later it becomes apparent that three of the six performers are in wheelchairs. In these opening moments of *Outside/In* the space around the dancers, which is limited by the closeness of the camera focusing on heads and shoulders, foregrounds certain aspects of identity; age, gender and 'race', while temporarily hiding another; the dancers' abilities.

The borders and limits of bodies and space come into contact in a very graphic way. Mouths, ears and noses which feature, are all bodily orifices, thus directing attention in an immediate way to the actual boundaries of the body, and the extent to which they are fluid and can be blurred and merged. The merging of bodies and space is emphasized in the choreography by the passing on from one dancer to the next of a 'piece of space' in the form of air which disappears into bodies through mouths and ears, for example. Mouths, lips, tongues, ears and noses are both inside and outside the body as is skin. Seeing edges of bodies and space as both inside and outside in this way allows for ambiguities and ambivalence; opening up rather than closing down possible readings. Playing with boundaries and border zones in this way is both actually transgressive and metaphorically so because of the different bodies involved and the intimacies exchanged. At these limits and extremes of bodies and space, at the edges and in the border zones, meanings, which contribute to ideas about subjectivity, are continually being negotiated.

This is why body/space relations are ripe for further investigation, because here there is potential for change and innovation and for rethinking subjectivity.

This analysis of the beginning of *Outside/In* has focused on actual bodies and the actual space between and around them, but it has also drawn on metaphorical ideas, associated with the fluidity of bodily and spatial borders and boundaries, which have implications for subjectivity, that can be explored further. In this book explorations of body/space relations veer between the actual and the metaphorical throughout, since they mutually inform each other. For example, in the first part of the book actual body/space relations are explored when dances are filmed in non-theatrical performance sites, such as beaches. In these spaces choreographic plays with spatial boundaries such as shorelines are examined to throw light on some of the ways in which metaphors such as borders and 'in-between' spaces work. The connotations that these different spaces suggest, and the resonances they have with these metaphors and ideas, can inform about how such spaces are constructed to have particular associations, how performers in them are also constructed, and how each contributes to the construction of the other. This is why most of the dances analyzed in this book are films or videos rather than live theatre performances, providing scope for a wide range of different spatial and environmental locations.

Space then, like subjectivity, is a construct, a human or social construct, and so it cannot be explored without reference to human subjects. Possibly the most immediate relationship of subjects to space is through their bodies since 'it is by means of the body that space is perceived, lived – and produced' (Lefebvre, 1991: 162). Constructions and conceptions of space produced in this way are inextricably bound up with conceptions of time such that 'it is not possible to disregard the fatal intersection of time with space' (Foucault, 1986: 22). Space as a product has a history (Lefebvre, 1991) which is forever changing through time. This is important when the roles of space and body/space relations in constructions of subjectivity in dance are examined. As was evident in the analysis of the opening moments of *Outside/In*, space and time are central to these processes of construction for the 'fundamental structuring categories [of space and time] have major consequences for . . . meaning and representation, subjectivity and agency, culture and society, identity and power' (Grossberg, 1996: 171). This is possibly also why ideas about space are currently prevalent in various strands of cultural theory.

Currency of ideas about space

The French poststructuralist, Michel Foucault's claim that 'the present epoch will perhaps above all be the epoch of space' (1986: 22) is at least partly borne out by the extent to which ideas and debates about space have surfaced in a range of disciplines. In relatively recent poststructuralist and postmodern theory there has been an almost obsessive interest in ideas and theories about space, the ways in which space is experienced and its characteristics as a social construct (for example, De Certeau, 1988; Foucault, 1986; Grosz, 1994b, 1995; Lefebvre, 1991). This interest has extended to several areas of cultural studies, including not surprisingly perhaps to geography. Important insights have been contributed by feminist geographers about the ways in which space is gendered, some of which are explored here in Chapter 2 (Massey, 1993, 1994; Rose, 1993 a,b). There has been a burgeoning of work employing spatial metaphors such as 'mapping' and 'cartography' as structural and cognitive tools particularly in the early days of postmodern theory (for example, Huyssen, 1984; Jameson, 1988, 1991).

In a broader sense, these spatial metaphors, such as travel, maps, mapping and cartography, have appeared in the titles of cultural theory articles, books and anthologies with considerable frequency (for example, Bell and Valentine, 1995; Bird *et al.*, 1993; Clifford, 1992; Devi, 1995; Diprose and Ferrell, 1991; Jarvis, 1998; Probyn, 1990; Zizek, 1994). The feminist cultural theorist Janet Wolff (1993), exploring the use of such metaphors, has indicated that they are gendered and in Chapter 2 of this book I apply her ideas to dance. However, despite the prevalence of interest in these debates about space in other artistic fields, such as visual art (Burgin, 1996; Florence, 1998; Lippard, 1997; Pollock, 1996), film (Burch, 1968; Deleuze, 1992; Jarvis, 1998; Mulvey, 1989), architecture (Colomina, 1992; Norberg-Schulz, 1971, 1980) and theatre (Aronson, 1981; Carlson, 1989; Chaudhuri, 1995; Read, 1993), where issues of the constructed nature of spaces and their effects have been investigated, the ideas have hardly been explored at all in dance theory – a few mentions in Thomas (1997) are the rare exceptions. Given the centrality of the body to dance, and the fact that the debates on and around space have extended to embrace relations between space and the body (for example, Pile, 1996, see also anthologies edited by Ainley, 1998; Duncan, 1996; Nast and Pile, 1998), dance should have much to contribute to this increasingly extensive area of current debate and research. In a very fundamental and immediate way dance presents representations of bodies in spaces, their relations to the spaces and to other bodies, and, as has been indicated, in this sense it is a most pertinent arena for exploring questions of subjectivity.

Questions of subjectivity

Notions of subjectivity are complex. What it means to be a subject entails being an 'I' a 'you' a 'he' or a 'she'. In other words subjects are constructed and positioned by and through language and discourse. Bodies or humans are subjected; they attain subject positions, or become subjects, in this way. Traditional Western notions of subjectivity position or fix the subject in time and space. Consequently different ways of conceptualizing time and space affect the ways subjectivity is understood. In the discourse of dance, when the limits of bodies, where they meet and interact with the surrounding space and with other bodies, are the focus, attention is drawn to the bodies' similarities and differences. Subjects are constituted as the same as others or different from others, through discourses such as dance. The similarities and differences of dancing bodies point to the kinds of subjects performing. They are seen as gendered, 'racialized', sexualized and in terms of their ability. Various historical, political, social, cultural, sexual and 'racial' discourses such as dance construct subjects to be seen in these ways. Yet despite these differences, subjects traditionally are seen as the same; as universally human and as unified. This is because in much Western philosophy subjection erases difference to maintain the illusion of sameness. Within the context of certain strands of postmodern theory the differences in subjects are recognized, so that subjects are seen as fragmented or split. This allows for subjects to be *both* the same *and* different from one another, to be *both* gendered *and* 'racialized', to be *both* agents with power to act on the one hand, *and* subjected to the rules and laws of language on the other. It allows for the occupation of more than one subject position. Seeing subjectivity as constructed in these ways importantly means it is open to change, it is in process, fluid and mobile rather than fixed.

This openness means that a representational practice like dance can call into question traditional notions of subjectivity as unified and fixed. This book, through a focus on space in dance, explores various ways in which dance can challenge, trouble and question these fixed perceptions of subjectivity. The questions it addresses include: how does the space in which the dance occurs affect perceptions of subjectivity? Different spaces for dance such as cities, and the buildings that constitute them, and wide open outdoor spaces, such as landscapes and beaches, hold connotations and associations. They are not empty. Like bodies, they can be gendered, 'racialized' and sexualized. What happens when dance is set in such places? What effect does this have on the choreography?; on the spaces?; on ideas concerning subjectivity? How does the space between dancers, and between dancers and spectators, affect constructions of subjectivity? How can investigations of body/space relations in dance contribute to rethinking notions of subjectivity, to opening up possibilities for previously excluded subjectivities?

Subjectivities have been excluded in the past because differences have not been recognized. Constituents of difference that make up subjectivity and identity, specifically gender, 'race', sexuality and ability, and how they are represented in postmodern dance, are a focus of my exploration of body/space relationships. My interest in these constituents of identity stems from a longstanding engagement with feminist theories, and more recently, with postcolonial and queer theories, all of which have emphasized the power differentials inherent in language and discourse which value one idea or concept over another resulting in discrimination and oppression. In each chapter selected examples of postmodern dance are read in the light of some of these theories in order to explore relationships between space and bodies in dance in ways that can aid an understanding of how identities and subjectivity are negotiated, constructed and resisted. In the process the power differentials at work are exposed, and ways of reading 'against the grain' in order to eliminate any potential for discriminatory or oppressive practices are suggested.

Reading dances

This book provides frameworks for exploring space and subjectivity in a range of selected examples of postmodern dance. Readings appro-priate to the choreography are posited but these are not intended as definitive. They do not exclude other readings, on the contrary, my explorations of the dances, read through particular theories, are intended to open up the dances for further readings. These explorations draw on a selective reading of pertinent feminist, postcolonial and queer theorists. They, in turn, draw on relevant phenomenological, philosophical, post-structuralist and postmodern theory. Post-structuralist and postmodern theories by unhinging notions of subjectivity and agency have been criticized for undermining the grounds of political action. I disagree with these criticisms. I am arguing throughout for political positions. I hope the book demonstrates that certain post-structural theories appropriately explored alongside radical postmodern choreography can demonstrate through their application the political potential of both the dance and of the theories in practice. In these readings I am not looking at space purely formally but rather examining the ideological, philosophical and political parameters of space. In other words I am concerned with spatial concepts such as perspective and distance and how these affect meanings in dance. In this sense I explore what and how space means in dance, how it is possible to think space differently, and what this means for dance and for subjectivity.

The key starting point is my interest in certain postmodern dances that raise questions and issues, concerning subjectivity and embodiment, which demand to be looked at and explored using a range of critical, and particularly spatial, theories. In the rest of this chapter I introduce the principal theoretical ideas that are explored in this book using incidents in Pina Bausch's film *Lament of the Empress* (1989) (*Lament* from now on) as empirical examples to illustrate them. Bausch is one of the most important postmodern choreographers of the last three decades of the twentieth century whose work, which is both arresting and contentious, has been performed and debated throughout Europe, North America and beyond. She confronts issues of subjectivity, identity and embodiment head on. The impact of her choreographic imagery is enhanced in many of her dances by the particular kinds of resonant spaces she creates on stage. In *Lament* Bausch exploits the increased scope the film medium gives her to explore this aspect of her work. Set in interior and exterior spaces in and around the German town of Wuppertal, where her dance company is based, *Lament* is made up of a series of seemingly unconnected closeups and long shots of individuals, performing a mixture of pedestrian and dance vocabulary in various bizarre costumes from swimsuits to ball gowns, occasionally in states of disarray. Performers are often seen struggling with the elements – snow, mud, wind, water. Sometimes they appear distressed or lost, crying out or shouting to attract attention, wandering aimlessly through woods, down roads or over hillsides. Surreal radical juxtapositions include a woman sitting in an armchair smoking at a busy traffic intersection, and a telephone mouthpiece held in a lavatory bowl while it is being flushed. Scenes shot in public spaces include passers by and onlookers. Many different urban, rural, indoor and outdoor locations are used as well as a range of indeterminate spaces between.

The various post-structuralist theories I explore when reading the dances may seem fragmented. However they are all concerned with the ways in which subjectivity is constructed in and by discourse, and with critiquing the premises of Western philosophy which revolve around the concept of an ideal, rational, unified, subject, which, in turn, relies on dualistic thinking that enforces seeing things in terms of binary oppositions. These premises result in constructions of subjectivity in choreography and performance, where the spectator is positioned conventionally as subject and the performer as object in particular relationships to space, time and discourse. Some of the implications of binary oppositions for constructions of subjectivity are explored immediately below under the heading, 'Bounded Bodies'. Following this, two interrelated ways of conceiving things differently are introduced which are explored throughout the book to problematize binaries – 'in-between spaces' and 'inside/outside interfaces'.

Bounded bodies

Since the French philosopher René Descartes (1596–1650), the subject has been thought of as situated in and positioning her/himself in space and time. As the American feminist theorist Susan Bordo observes, Descartes conceptualized the body 'as the site of epistemological limitation, as that which fixes the knower in time and space and therefore situates and relativizes perception and thought' (1995: 227). It is within this context of body/space relations that subjectivity and subjects' relations to the world are understood. The rational unitary subject, inherited from Descartes, is reduced to finite co-ordinates in time and space, where time and space are seen as unproblematic, quantifiable and measurable in a scientific and mathematical way. By bracketing off other ways of experiencing time/space, the subject is confined to a private time/space whose isolation from the private times/spaces of others ensures the disinterestedness and distance of individual judgement. The ways in which the subject is constructed as distanced from the world, the separation of subject from object and the forces or technologies of power at work particularly in relation to the body in this construction process are theorized by Foucault, whose ideas inform the analysis in Chapter 7.

One of the aims of this book is to show how postmodern dance can challenge these ideas of subject/object separation. For example, there are moments in Bausch's *Lament of the Empress* when ordinary people are seen watching the action. There is a scene when a woman looks out from her apartment window, later a net curtain is drawn obscuring another onlooker from view. This challenges the conventional positioning of the spectator as subject and the performer as object, because the woman looking through her window – an 'innocent bystander' – becomes the *object* of attention when filmed. The conventional subject/object separation and fixity is disrupted. The normal relation of subject/object paralleling spectator/performer is blurred, as is the relationship between private and public space. For a moment in the film the private space of the woman's apartment becomes public as it is filmed, but then the drawing of the net curtain reminds the spectator that this is a private space where the camera would not normally intrude.

This subject/object separation is part of a whole series of associated dualisms or binaries such as self/other, mind/body, outside/inside, male/female which are hierarchized. The first term in the pair is valued and privileged over the second which is seen as the first's 'suppressed, subordinated, negative counterpart' (Grosz, 1994b: 3). This is illustrated in the excerpt from *Lament*; the woman observing from her window, who as an unseen spectator was in the privileged position of subject, becomes more vulnerable as an object of the camera's attention. Importantly these binaries are spatially constructed and as evident in the representational practice of dance as in any other cultural practice. As Australian feminist theorist Elizabeth Grosz suggests, it is not the pair or the number two that is the problem: 'rather it is the one which makes it problematic'. She continues, 'the one in order to be a one must draw a barrier or boundary round itself'. This implicates it 'in the establishment of a binary – inside/outside, presence/absence' (ibid: 211). This kind of corporeal spatiality where bodies are seen and conceived as bounded entities is central to understanding subjectivity from a Cartesian perspective. The bounded body is evident in much Western theatre dance such as classical ballet, where bodily containment and control are paramount. Throughout this book many of the postmodern dances discussed play with and blur the boundaries of the body in different ways disrupting, and challenging the fixity of identities seen in the context of binaries.

The traditional relations of bodies in space, which depend on a separation of subject and object and on a dualistic way of seeing the world, crucially also infer at least two other interrelated concepts, which construct space and bodies in particular ways, and which are investigated here in dance. One is the notion of *perspective*, bound up with ideas about visualization and the dominance of the visual in Western culture. The other, linked to conceptions of perspective and its implication of direction towards a vanishing point, void or lack, is *desire*, which is traditionally generated, at least in part, because of the separation of subject and object, self and other.

Perspective and vision

The notion of perspective, which emerged during the Renaissance but developed from Ancient Greek Euclidean geometry, is bound up with the construction of a particular kind of subjectivity or subject position. It implies a single unifying viewpoint from which and to which all points converge, evident in dance in the traditional performer/spectator relationship. Discussing this in the context of the French psychoanalyst Jacques Lacan's influential theories about the construction of subjectivity, Grosz argues:

> for the subject to take up a position as a subject, it must be able to be situated in the space occupied by its body. This anchoring of subjectivity in its body is the condition of coherent identity, and moreover, the condition under which the subject has a perspective on the world, and becomes a source for vision, a point from which vision emanates and to which light is focussed.
>
> (1994b: 47)

As the French post-Marxist theorist Henri Lefebvre (1991) has revealed, visualization and perspective are deeply pervasive ideological constructs, which often remain invisible, but in fact structure the ways in which space, bodies and experience are perceived and understood in a limiting and reductive manner.

One of the purposes of this book is to identify these constructs in dance and expose the ways in which they shape experience. Early in the film of *Lament* a woman dressed as a Hugh Heffner 'bunny girl' is seen struggling and stumbling across a very muddy hillside, her costume is in disarray, her breasts are half-exposed and in each hand she carries a high-heeled shoe. It is a bizarre image. Much later in the film the same bunny girl is seen, still in disarray but this time wearing her shoes, running towards the camera from the distance down a very long narrow country road. There is nothing else in the frame, simply the long road with empty fields to either side, and the girl. The shot constructs the traditional single viewpoint of perspective to which and from which lines of vision converge, by having a single figure – the bunny girl – on the long road. If the bunny girl had been a regular Heffner girl with costume intact, light emanating from its satin surfaces, filmed in her normal glamorous habitat of a night club, then the single viewpoint of perspective would be reducing and limiting her to an objectified sex object for men's pleasure. This is how perspective normally works. However, because in Bausch's film she is muddy and in disarray, the single reading of what or who she is, is disrupted; at the same time, placing her on a long road which disappears into the distance, reminds viewers of the way in which perspective normally works to focus vision.

The importance of the visual in theories which construct bodies and space cannot be underestimated. 'The vanishing-point and the meeting of parallel lines "at infinity" were the determinants of a representation, at once intellectual and visual, which promoted the primacy of the gaze in a kind of "logic of visualization"' (Lefebvre, 1991: 41). The gaze, arising out of the logic of visualization implied by perspective, constructs and is constructed by a rational unified subject who uses it to control and objectify any 'other'. Using Picasso as an example, Lefebvre links this notion of looking or visualization to violence, a particular kind of masculine violence. He suggests: 'Picasso's cruelty towards the body, particularly the female body ... is dictated by the dominant form of space, by the eye and by the phallus – in short, by violence' (1991: 302). The British feminist film theorist Laura Mulvey's (1975) application of the male gaze to film, although drawing more from psychoanalytic theory, develops a similar concept which has been much used and adapted in feminist work. The appearance of exhaustion, disarray and distress of the bunny girl in *Lament* which renders her vulnerable, both actually and metaphorically can be seen as the result of violence – the cumulative effect of years of infliction of the male gaze.

The ways in which these concepts have become powerfully and deeply ideological are evident in Lefebvre's claim concerning perspective that, 'at all levels, from family dwellings to monumental edifices ... the elements of this space were disposed and composed in a manner at once familiar ... which even in the late twentieth century has not lost its charm' (1991: 47). Lefebvre continues stating that the 'spatial code' of perspective 'is not simply a means of reading or interpreting the space: rather it is a means of living in that space, of understanding it and of producing it' (ibid: 47–8). Lefebvre claims these ideological views are also reductive, limiting the imagination and implying closure, and that this is only too evident in the practice of architecture (and I would add dance) where, he argues, the space of the architect is 'a visual space, a space reduced to blueprints, to mere images – to that "world of the image" which is the enemy of the imagination. These reductions are accentuated and justified by the rule of linear perspective' (1991: 361). These claims are explored alongside dance in the investigations of body/space relations that follow. Specifically in Chapter 10, links between architecture and dance in terms of body/space relations are explored to expose the limitations of the visualization process, and to explore ways of seeing differently and going beyond the reductive single viewpoint of perspective.

Desire

Desire relates to perspective in the sense that it is also traditionally dependent on a vanishing point for its comprehension. It conventionally relies on the separation of subject and object and the distance between them, which also depend on a notion of space as empty or lacking. In this sense 'space . . . unleashes desire. It presents desire with a "transparency" which encourages it to surge forth in an attempt to lay claim to an apparently clear field' (Lefebvre, 1991: 97). The constructed nature of 'transparency', which Lefebvre puts in inverted commas here, and the use of the word 'apparently' are important, for they suggest that space is not empty or transparent, but for certain purposes it can appear so. Lefebvre terms this the 'illusion of transparency' (ibid: 27–8) because he claims that the socially constructed nature of space is in part concealed by this illusion. Some of the ways in which space is seen as transparent in this sense and linked to a masculine gaze which genders space are explored in Chapter 2. The ways in which subject and object, self and other, performers and audience and dancers and other dancers are thought, conceived of and seen as separate and distanced, depend on this view of space as empty or 'transparent'. This in turn 'unleashes desire' in a traditional sense which, linked to the ideas about perspective and the single masculine viewpoint cited above, can also be seen as limiting and reductive.

These operations of desire are directly confronted and challenged by the image of the muddy bunny girl in *Lament*. The ways in which the mud on the bunny girl link her to the surrounding hillside initially and then later in the film to the country road stretching between muddy fields, displace her from her metaphorical pedestal where she appears separate, in a clear field, as an object of desire for the masculine gaze. The mud and the countryside space have interrupted or got in the way of the distance necessary for desire to operate in the traditional spatial sense described above. There are many images that disrupt the conventional spatial workings of desire in this way in *Lament*. Throughout most of the film women and men are seen in vulnerable and often messy circumstances which almost always directly connect them to the particular environment in which they are performing, getting in between them and the spectator's gaze. The gap required for desire to operate in a conventional manner is repeatedly traversed messily linking the performers graphically with their surroundings. A scantily dressed woman dances in the midst of a blizzard, a man is seen buried under a heap of snow, a woman, one arm out of the sleeve of her dress, wanders as if lost through a wood of tall trees. These images and many more in the film serve to disrupt and challenge the conventional workings of desire, which depend on a violent masculine gaze, that in turn relies on perspective and the logic of visualization. As becomes evident in Chapter 2 however, simply presenting a messy body that connects with its surrounding environment, is not sufficient to disrupt the conventional workings of desire. The choreography and, where relevant, the filming of the body also play an important role.

Work in phenomenology and post-structuralism has suggested alternatives to these limited ways of viewing things. Instead of seeing things in terms of separation, of opposites, and from one perspective only, various theorists have explored the possibilities of focusing on the spaces in between; on spaces of ambiguity and hybridity, and of becoming, rather than being.

In-between spaces

Notions of in-between spaces are deployed on various levels throughout this book because they can problematize, challenge and offer an alternative to the dichotomies of binary oppositions. Bodies and subjects can be considered to be 'in-between' because of a range of ambivalences that are inherent in their construction. These ambivalences can be particularly relevant for the problematization of binary oppositions. Grosz proposes that:

> these pairs [binaries] can be more readily problematized by regarding the body as the threshold or borderline concept that hovers perilously and undecidably at the pivotal point of binary pairs. The body is neither – whilst also being both – the private or the public, self or other, natural or cultural, psychical or social . . . This indeterminable position enables it to be used as a particularly powerful strategic term to upset the frameworks by which these binary pairs are considered.
> (1994b: 23–4)

The bodies and the spaces in *Lament* are at times ambivalent and blurred in this way. They are not private or public, but both. The 'natural' and the 'cultural', the 'psychical' and the 'social' overlap and blur because of the radical juxtapositions created. The bunny girl who stumbles around on a muddy hillside is an ambivalent body in this sense, as is a man who attempts assiduously to shave crouching in a wet gutter, getting soaked by the spray of passing cars. These images transgress and blur boundaries because the characters are out of their 'normal' place. They create and play in spaces in between. The bodies act as 'borderline concepts' in Grosz's terms.

The potential of in-between bodies and in-between spaces, which exist at borders and in frontier zones, has been examined by, among others, postcolonial theorists, Homi Bhabha and Paul Gilroy. Bhabha, whose theories are discussed in Chapter 6, explores the ways in which concepts of the 'in-between' in terms of spaces and subjects can be mobilized to rethink identity. He argues:

> hybrid hyphenizations . . . emphasize the incommensurable elements as the basis of cultural identities. What is at issue is the performative nature of differential identities: the regulation and negotiation of those spaces that are continually, contingently, 'opening out', remaking the boundaries, exposing the limits of any claim to a singular or autonomous sign of difference – be it class, gender or race . . . difference is neither One nor the Other, but something else besides – in-between.
>
> (1994: 219)

Examples of cross-dressing in *Lament* – various different male characters wear evening dresses – could be seen as 'hybrid hyphenizations' which 'emphasize . . . incommensurable elements as the basis of cultural identities'. They remind viewers that gender identities are *performed*, and by performing them otherwise, open out the spaces in-between.

Paul Gilroy also recognizes the distinctively fluid characteristics of in-between spaces in his writings about the African diasporic space that he terms 'the black Atlantic'. He argues:

> the black Atlantic provides an invitation to move into the contested spaces between the local and the global. . . . The concept of space itself is transformed when it is seen less through outmoded notions of fixity and place and more in terms of the ex-centric communicative

circuitry that has enabled dispersed populations to converse, interact and even synchronise.

(1996: 22)

In-between spaces in these terms, which are seen as animated by diverse marginal voices that find points of local/global connection because of shared experiences of displacement, are explored with reference to Gilroy's theories in Chapter 7. In *Lament* 'contested spaces between the local and the global' are foregrounded by Bausch's broad choice of music from different cultures for the soundtrack. Some examples are Italian folk music, North American Indian dance music, Hungarian Czardas, Billie Holliday's *Strange Fruit* and Argentinian dance music. This range of music from different cultural sources, together with the performances of the dancers, constructs the spaces, in and around Wuppertal, where the dancers are performing, differently. They have echoes of Gilroy's 'ex-centric marginal spaces', in-between spaces, on the borderlines. Post-colonial theories such as these are discussed in Chapters 6 and 7 alongside postmodern dances that explore 'racial' aspects of identity.

Feminist theorists have also employed notions of the 'in-between'. The Italian-born Rosi Braidotti, currently based in the Netherlands, whose theories of nomadic subjectivity are explored in Chapter 6, stresses transmobility and the ability to move across and between boundaries in an interconnected way. She writes of 'in-between spaces where new forms of political subjectivity can be explored' (1994: 7). The American Donna Haraway, whose theories inform Chapters 2 and 4, also develops ideas of hybridization related to notions of the in-between in her concepts of cyborgs as futuristic machine/organism fabrications for a post-gender world. Emphases on permeable boundaries and affinities and networks in her theories ally them with the transmobility and interconnectedness of Braidotti's nomadism. In *Lament* connections and affinities between the disparate characters portrayed are suggested by various devices such as the soundtrack, where one piece of music often accompanies several contrasting adjacent scenes linking them across and between boundaries.

Grosz is more concerned with the ways in which the in-between can be applied to bodies, particularly female bodies, in an exploration of their permeability. She suggests that 'womanhood' occupies a kind of 'in-between' space. Discussing menstruation, she claims that it 'marks womanhood ... as outside itself, outside its time ... and place ... and thus a paradoxical entity, on the very border between infancy and adulthood, nature and culture, subject and object, rational being and irrational animal' (1994b: 205). The use of fluidity in general, and bodily fluids in particular, as metaphors for the mobility and ambiguity of in-between bodies and spaces, recurs in a range of different but related theories, where they are employed to further disrupt binary thinking. The French-based Bulgarian feminist Julia Kristeva argues that excessive overflows of bodily fluids and waste, which she terms abject, can suggest the merging of subject and object that results in an ambiguous subject in process. In *Lament* a woman expresses milk from her breast and drinks it. This subversive action plays with and blurs inside/outside bodily interfaces and boundaries. The use of actual bodily substances in dance can sometimes point to the same or similar meanings that are suggested metaphorically in these theories. Examples of postmodern dances, which feature bodily flows that have the potential for inhabiting the in-between spaces and blurring inside/outside body boundaries, are explored with reference to these theories, specifically in Chapter 8, but also in Chapter 9.

Inside/outside interfaces

As was evident in the discussion of *Outside/In*, and the example of the woman drinking her own breast milk in *Lament,* the boundaries or borders of the body are fluid and permeable. Borders can be seen to contain, surround and delimit movement in space, but they can also be crossed and broken. The potential of bodies to overflow or go beyond their conventional borders in dance, and to reveal ways in which body boundaries can be blurred and exceeded both actually and metaphorically, is a theme that recurs throughout the book.

Ideas about excessive overflowing bodies are explored in detail by the Russian literary critic Mikhail Bakhtin in his studies of medieval carnival (1984) and his development of the notion of 'grotesque realism', which are the focus of Chapter 9. He claims, 'the grotesque image displays not only the outward but also the inner features of the body: blood, bowels, heart and other organs. The outward and inward features are often merged into one', in this sense, he continues, 'grotesque imagery constructs what we might call a double body' (1984: 318). The examples, already cited, of messy, wet, muddy, snow covered bodies, often exceeding their costumes in *Lament,* along with the woman drinking her breast milk, can be seen as grotesque and 'double bodies' in Bakhtinian terms. They all play with and blur the inside/outside bodily interfaces and boundaries.

Bhabha, whose ideas of postcolonial hybridity evoke spaces of ambiguity where boundaries between inside and outside are blurred, quotes Bakhtin, who employs the term 'hybrid' to describe his 'double body'.

This is because it is 'double-voiced', 'double-accented' and 'double-languaged' presenting 'the collision between differing points of view on the world' (quoted in Bhabha, 1998: 33). Bhabha is quick to stress that the doubleness of hybridity, because of the idea of collision inherent in the concept, is very different from the doubleness of binary thinking, characterized by separation and opposition. In *Lament* the image of the muddy, exhausted bunny girl aligns bunny girls with real bunnies who burrow in the earth and mud, thus ridiculing the original, to which Bausch is referring. However, at the same time, the sheer exhaustion, distress and disarray of the girl evokes the demoralizing, dirty and sordid nature of the sex business that created her. This 'double-accented' image is both powerful and grotesque, suggesting what Bakhtin calls 'potential for new world views' (ibid: 33).

The close relations between the inner and outer limits of Bakhtin's grotesque body result in its 'open unfinished nature, its interaction with the world' (Bakhtin, 1984: 281). Throughout *Lament* Bausch's performers often display open, unfinished, messy bodies that closely 'interact with the world'. The radical juxtapositions inherent suggest ways of viewing subjects otherwise, of thinking them differently, of opening things up rather than closing them down. The unfinished nature of these 'grotesque' bodies means that, they are, as Bakhtin argues, bodies 'in the act of becoming' (1984: 317). This notion of bodies seen as 'becoming' and always in the process of constructing and being constructed – never finished – is one that recurs throughout the book. It is of particular use because it recognizes the non-fixity and instability of subjectivity, such that the subject never reaches a stable state of being which can be fixed in a binary opposition, it rather has the possibility of fluctuating in the spaces in between.

The French post-structuralists Gilles Deleuze and Felix Guattari, employ the notion of 'becoming' to characterize processes of continual transformation. This idea can be explored in relation to not only the space between bodies, but also to linkages and interfaces within and between bodies. These notions are bound up with ideas of multiplicity and productivity. Their theories of assemblages which focus on dynamic processes of interconnectivity characterized by flows, intensities and linkages, suggest further ways of theorizing inside/outside interfaces of bodies and spaces which, following Grosz's suggestions for refiguring desire are explored in Chapter 5. The dynamism of these ideas which continually stress ongoing processes rather than fixed products makes them particularly attractive for dance. The evocative imagery of Baroque folds and foldings has inspired Deleuze to explore related notions of infinite, wave-like energies. The theories that resulted are discussed alongside dance in Chapter 8.

Many of these ideas concerned with inside/outside bodily interfaces owe a debt to the French phenomenologist, Maurice Merleau-Ponty. His notion of a 'double sensation' suggests ways of thinking beyond the limitations of dualism and the bounded entities that result. Grosz suggests it 'creates a kind of interface of the inside and the outside' (1994b: 36). She explains this, stating, 'the information provided by the surface of the skin is both endogenous and exogenous, active and passive, receptive and expressive'. She continues, 'double sensations are those in which the subject utilizes one part of the body to touch another, thus exhibiting the interchangeability of active and passive sensations, of those positions of subject and object, mind and body' (ibid: 35–6). These notions are similar to the Belgian born French-based feminist Luce Irigaray's image of the touching of two lips in her description of female sexuality. However, in her focus on sexuality she departs fundamentally from Merleau-Ponty who Irigaray criticizes for failing to take sexual difference into account (Irigaray, 1993: 151–84). The importance of touching as a way of filling the traditional gap associated with an unproductive notion of desire, is explored with reference to Irigaray's theories in Chapter 5.

Spatial interfaces do not exhibit the same physical features as bodily interfaces, but their inside/outside boundaries can also be seen to touch or merge. In *Lament* when a woman is seen sitting alone in a constantly circling monorail car uttering a drunken monologue, the public space of the monorail car is rendered private and intimate by the woman's performance. The private/public boundary of this space blurs in the performance. The monorail car is an interior space, but in its circling it is connected to the exterior spaces of Wuppertal, actually by the monorail and virtually by the views of Wuppertal seen through its windows. It also connects the woman to the city as it moves her through it. The boundaries of the city spaces and of the monorail car space touch and merge. As well as disrupting the logic of visualization derived from a single perspective viewpoint, this scene also highlights the contingency and the effect of spatial boundaries. Similar examples of blurred spatial boundaries, often involving buildings and dancers, are explored in Chapters 3, 6 and 10.

To summarize: when examining dance focusing on body/space relations, attention is drawn to the boundaries of bodies and space, where the limits of representations of embodied subjectivity are forged, and become apparent. This is why such a focus is particularly pertinent when constructions of subjectivity are at issue. This focus on body/space relations prompts investigation of a set of interrelated ideas current in contemporary critical theory, all concerned to critique the premises of Western philosophy, which result in seeing things in terms of binary oppositions. These interrelated ideas surface and resurface in different forms in different chapters. Their application to the analyses of certain postmodern dances which raise issues of subjectivity, enables explorations of the construction of subjectivity such that power differentials are exposed, and suggestions for reading 'against the grain' can be made. These key interrelated ideas, current in critical theory, are the blurring of bodily and spatial boundaries, the in-between spaces of hybridity and ambiguity, a focus on inside/outside interfaces, privileging touch and sensation over the visual, focusing on foldings of bodies and space, and on bodily and spatial excesses. When dance is examined using such ideas, the mutual construction of bodies and spaces is at the centre of the investigation. This notion, inherent in Lefebvre's theories, is developed by Grosz from ideas concerning bodily inscription derived from Foucault. This focus on the physical and metaphorical construction of bodies and spaces, and their interactions, directs attention to the subtleties and nuances of meaning inherent in those postmodern dances that address, implicitly and explicitly, issues of embodied subjectivity. Consequently, ways of rethinking and challenging traditional notions of subjectivity to make room for subject positions that are currently excluded become possible.

The book is a collection of close readings of postmodern dances. The dances, because of their intricacies and complexities, have directed me to certain post-structuralist theories to aid my understanding. Engaging with the dances in this way has illuminated aspects of the theories. Dance with its focus on the body asks particular questions of theory, it teases out particular emphases in theory often in refreshing new ways and it can illustrate some of the more abstract ideas physically and immediately in an embodied manner. For example, issues of gender, which may be implicit in the theories, can become explicit when examined in dance. In turn, the theories have directed my viewing of the dances to a certain extent. They have brought insights to the choreography revealing previously hidden meanings. Subversive nuances and subtleties in the dances have been brought to light because of the focuses of the theories applied. The book, therefore, demonstrates the mutually beneficial reciprocity of dance and theory.

These then constitute the ideas and theories behind this book. Each is discussed, explored and expanded in relation to specific postmodern dances. The theories outlined are not discussed or explored in any logical, linear or indeed chronological manner. Rather they 'come and go' throughout the text in a fluid fashion deemed appropriate given their non-linear character and notions of interconnectivity which they embrace. To a certain extent, chapters and parts of the book stand alone and there is no one recommended way of reading. The book is loosely organized in three parts: 'Constructions of space and subjectivity', 'Dancing in the in-between spaces', and 'Inside/outside bodies and spaces', with a final chapter that shows how the concerns of all three parts can overlap. In the first part, site specific dances are examined focusing on the ways in which the actual spaces of their location – the particular places in which they are set – are constructed, and in turn contribute to constructions of subjectivity. In Part II there is a shift from these specific places – these actual spaces – to virtual or metaphorical spaces in the dances seen as in-between. These are created as in-between by the choreography and, in some cases, the filming. They are ambivalent and indefinite, neither one thing nor the other. Consequently they suggest possibilities for rethinking and challenging constructions of subjectivity as fixed. In Part III the focus shifts to the dancing bodies, specifically to the inside/outside spaces of their borders and boundaries, and the ways in which these can trouble and subvert traditional constructions of subjectivity and make room for previously excluded subjectivities. The final chapter illustrates that these focuses are not mutually exclusive, but overlapping; they interconnect in various ways which are apparent throughout the book. The material for some of these chapters started life as chapters of other books and conference papers published elsewhere, but in bringing it together and reworking it, key arguments that cohere around a set of interrelated ideas about bodies, space and subjectivity are developed, which call into question traditional notions of subjectivity and suggest possibilities for rethinking it.

In the first part of the book entitled 'Constructions of space and subjectivity' different ways in which spaces and dancing bodies can be seen to construct each other and the implications for subjectivity are examined. The first chapter focuses on four site specific dance films by European and British choreographers and directors, which all focus on travel of one form or another and are set in different spatial locations. It explores the extent to which the gendering of the spaces is hidden or revealed by the choreography, performance and filming, together with the effects of this on the construction of subjectivity and the power differentials at work. Chapter 3, after focusing on some early American postmodern work set in and around the city of New York which provides precedents, examines two dance films, by British and European choreographers and directors, set in cities; the first, in a Brussels alleyway, the second, in a flat in London. This chapter focuses on the ways in which the dancing bodies and the city spaces mutually define each other, and how identities of bodies and spaces are invested with power of different kinds deriving from gender. A beach hut is the site of the British choreographer, Lea Anderson's *Out on the Windy Beach* (1998) analyzed in Chapter 4. Associations of seaside resorts with leisure and the erotic on the one hand, and the mythic, dark, environmental force of the sea on the other, are explored in the choreography, to reveal the complexity and fluidity of mutual constructions of space and subjectivity. Focuses on coastal borders in the choreography are explored literally in the form of plays with shorelines, and metaphorically through surreal and ironic choreographic elements which are shown to subversively point to new possibilities for subjectivity.

In Part II entitled 'Dancing in the in-between spaces' the focus is less to do with the spaces of particular locations, and more to do with actual and virtual 'in-between' spaces. In Chapter 5 the ways in which desire, specifically lesbian desire, fills, plays in and occupies the spaces between dancers and between dancers and spectators, is the focus. The chapter argues that lesbian desire is spatialized differently and this is demonstrated through analysis of four postmodern dances that can be read as lesbian, choreographed by British and European choreographers. It is suggested that distances and differences between subject and object, and between self and other are minimized, because these actual and metaphorical boundaries in the dances are blurred and fluid. The next chapter examines British-based Shobana Jeyasingh's *Duets with Automobiles* (1993) to explore the in-between spaces of hybridity and nomadic subjectivity negotiated in the piece by three classical Indian dancers performing in empty London office buildings. The possibilities for rethinking questions of meaning, representation, identity and female subjectivity in terms of blurred boundaries and interconnectedness are revealed. The actual space in between the dancers and the architecture is also explored for its potential to suggest new identities. The final chapter in this section is concerned with the diasporic in-between spaces that feature in works by the American postmodern choreographer, Meredith Monk, and the British rap artist, Jonzi D, where issues of 'racial' identity, difference and displacement are addressed. The works explore ways in which systems of power operate spatially to construct subjects, but also how these can be resisted through dance to reveal the possibilities for constructing fluid identities and subjectivities that celebrate difference.

Part III entitled 'Inside/outside bodies and spaces', focuses on the borders and edges of bodies and spaces where inside and outside exist in an ambivalent relationship of doubleness. The possibilities of this liminal and libidinal space for rethinking embodied subjectivity are examined. Chapter 8 explores these fleshy borders in three very different works: Trisha Brown's *If you couldn't see me* (1994) from America, Lea Anderson's *Joan* (1994) and Yolande Snaith's *Blind Faith* (1998) from Britain. It examines the flesh, fluids and folds in the unfamiliar territory of a woman's back (*If you couldn't see me*), in the actual and metaphorical inside/outside bodily forms inspired by imagery from Renaissance and Baroque paintings and sculptures (*Blind Faith*), and in the actual and spiritual spaces of the 'internal choreography' of the mythical martyr, Joan of Arc (*Joan*). The next chapter explores the extent to which the excessive and grotesque bodies of Liz Aggiss' *Grotesque Dancer* (1987 revived 1998) from Britain and Mark Morris' *Dogtown* (1983) from America, thrown up by the 'troubled' spaces of modern and postmodern urbanity, can be subversive in a Bakhtinian sense. This analysis is juxtaposed with that of Emilyn Claid's *Across Your Heart* (1997) for CandoCo Dance Company from Britain, where the polarizing of classical and 'grotesque' bodies in Western culture is used to explore the potential for thinking disability and sexuality otherwise.

By way of conclusion the final chapter examines the dance film, *Rosas Danst Rosas* (1997) choreographed by the Belgian choreographer, Anne Teresa De Keersmaeker, and *Enemy in the Figure* (1989) choreographed by the American Director of Ballett Frankfurt, William Forsythe. It is argued that the dances, which are both concerned with architectural spaces, deconstruct space by disrupting the 'logic of visualization' and point to ways of experiencing space differently which affect subjectivity. The chapter provides a focus for revisiting the bodily and spatial themes and ideas explored in the book and reiterating the interrelations between them and their potential for suggesting possibilities for rethinking subjectivity.

Part I
Constructions of Space and Subjectivity

2
Travel Metaphors in Dance – Gendered Constructions of Travel, Spaces and Subjects

Introduction

In the next three chapters, dances set in mainly outdoor spaces are examined, focusing on the ways in which the physical spaces of their location can be seen to be constructed in part by the dances, and in turn can contribute to constructions of the dancing subjects. The role gender plays in these constructions is a focus.

This chapter explores the treatment of travel, space and subjects in four site specific dance films and videos which focus on travel of various kinds. The ways in which travel is portrayed are bound up with the locations used, the particular places in which the dances are set, and how these spaces are constructed. The interest in travel as subject matter for dance parallels the use of travel metaphors in recent postmodern theory. The crisis in subjectivity inherent in postmodernity has been described in terms of metaphors associated with travel such as nomadism. The complexities of the crisis evident in fragmented subjectivity and in changing conceptions and experiences of space, it has been suggested, require navigation or mapping. It has been argued that these travel metaphors, like real travel, are gendered (Wolff, 1993). It has also been argued that social constructions of space are gendered (Rose, 1993a,b). The ways in which constructions of travel, of the spaces of travel and of the subjects that travel in them, are gendered, are identified in the dances as either hidden or revealed. I argue that revealing the ways in which travel, space and subjects are gendered can point to new ways of conceiving subjectivity.[1]

After briefly outlining some uses of travel metaphors in postmodern discourse and showing how travel and space are gendered, examples of the gendering of travel, spaces and subjects are identified in the dance films. In conclusion suggestions are made for reappropriation of the travel metaphors and new ways of conceiving subjectivity.

Metaphors of travel in postmodern discourse

New experiences of space and time are claimed to be characteristic of the present postmodern era (Bird *et al.*, 1993). The concept 'time–space compression' based on the notion that distances in time and space are shrinking is explored in the geographer David Harvey's *The Condition of Postmodernity* (1990). The world, through new technologies, is in many senses a much smaller, more accessible place than it used to be. Travelling of all kinds, of information as well as people, through time as well as space, is escalating. Whereas the motorcar, the train and the aeroplane could be seen to be signs of modernity, postmodernity is rather characterized by a range of electronic, computerized, telecommunication, video and virtual reality developments.[2]

These developments have revealed tensions between 'the global' and 'the local', between 'space' and 'place'. Attempts to express, understand or resolve these tensions are evident in a nostalgic fascination with local cultures and the vernacular or the 'other' in some postmodern culture. 'Travel' as a metaphor gives access to the 'other'. Metaphors of travel position the 'other' as either 'natural' or constructed and in doing so they either hide or reveal key constituents of the construction, such as gender.

While on the one hand the links between different cultures, times and places stress integration or a kind of globalization, in another sense recognition of the range of cultural groups and 'others' that exist, emphasizes notions of fragmentation or localization, which are also manifest in concepts of a 'fragmented subject'. In order to address the fragmentation of not only subjects and identities, but also of concepts of 'meaning' and 'truth', postmodern discourse has employed metaphors of travel. In a text, tellingly titled *Cartographies – poststructuralism and the mapping of bodies and spaces* (Diprose and Ferrell, 1991), it is suggested that 'the philosopher of the future will be a "wanderer", a transient who abjures attachments to existing institutions and ideas... the desert in which he or she wanders is... a mobile space characterized by the variable directions and the multiple dimensions in which movement is possible' (Patton, 1991: 53).

Culture can be rethought in terms of travel to accommodate the shifting 'truths', encounters, horizons and landscapes of postmodernity so that culture as a 'rooted body that grows lives and dies is questioned' and replaced by 'constructed and disputed *historicities*, sites of displacement, interference, and interaction' (Clifford, 1992: 101). Recognition that the terrain is varied and complex results in an extension of these travel metaphors to include the notion of worldviews becoming 'maps', 'topographies' or grids with 'coordinates' as reference points. The notion of 'cognitive mapping' as a postmodern way of 'seeing the world' has been championed by the American Marxist critic, Frederic Jameson (1988) and the term 'conceptual map' (Connor, 1989) has also been used. As indicated in the introduction to this volume, several postmodern texts have emerged with 'mapping' in their titles. Given that everything, from identities to ideas to cultures characterized as postmodern, is essentially fluid, forever shifting and changing, on the move, it is perhaps not surprising that travel metaphors of various kinds have become current in postmodern discourse.

Travel and gender

The ways in which travel is gendered often remain hidden. Some of the theorists who use travel metaphors, particularly Harvey (1990) and Jameson (1991), almost entirely ignore gender in their analyses (Massey, 1993, Morris, 1993). Arguing that travel metaphors in cultural criticism are gendered, the feminist cultural theorist, Janet Wolff, claims that 'just as the practices and ideologies of *actual* travel operate to exclude or pathologize women, so the use of that vocabulary as metaphor necessarily produces androcentric tendencies in theory' (1993: 224).

Having examined vocabularies of travel in postmodern theory, postcolonial criticism and post-structuralism, Wolff looks at women and travel and concludes that 'histories of travel make it clear that women have never had the same access to the road as men' (1993: 229). She quotes the feminist historian, Cynthia Enloe, who argues 'being feminine has been defined as sticking close to home. Masculinity by contrast has been the passport for travel' (ibid: 229). Wolff points out that women *do* travel and have a place in travel, but often that place is marginalized and degraded, as in tourism where women are often hotel maids or active in sex tourism. Those women who have travelled in the past have often been masculinized and seen as eccentric for taking up a 'male pursuit'. Wolff cites the hostile criticism of the film *Thelma and Louise* (1991) as partly to do with the unacceptable notion of women starring in a road movie, normally the domain of men.

'Thelma' and 'Louise' become masculinized as the film progresses as they move from the 'supposedly female space of the home to the supposedly male space of the great outdoors' and take to the road which acts as a sign for 'a certain mythicized freedom' (Tasker, 1993: 136). It is claimed that '"Good travel" (heroic, educational, scientific, adventurous, ennobling) is something men (should) do. Women are impeded from serious travel' (Clifford, 1992: 105). The journalist, Dea Birkett, reviewing books on women travellers in the last two centuries, comments, 'it is far ... more demanding for a woman to wander now than ever before' (1990: 41).

Space, power and gender

Space as a concept associated with travel can also be seen to be gendered. As a social construct space is not transparent and innocent, it is imbued with power of different kinds. In this sense there is a *politics* of space and of travel as movement through space. For example, the forced migration of people in the face of lack of resources or employment and the regulation associated with refugee camps and immigration, demonstrate the *powerlessness* of some kinds of travel. The feminist geographer, Doreen Massey employs the term 'power-geometry' to describe the differential power associated with movement flows and travel. She compares those who are doing the moving and communicating and importantly are in control of these processes – the 'jet-setters' – and those who, although moving a lot, are not in charge in the same way – the refugees, migrant workers, third world peasants, and so on. As she indicates, 'mobility and control over mobility both reflect and reinforce power' (1993: 62). This power differential is based to a certain extent on power structures and institutions that operate elsewhere, such as patriarchy.

The feminist geographer, Gillian Rose, discussing the ways most geographers look at space employing what she terms 'the geographer's gaze', which is bound up with their claim to knowledge, argues that it 'rests on a notion of space as completely transparent, unmediated and therefore utterly knowable' (1993a: 70). She cites the French, post-Marxist theorist Henri Lefebvre who proposes that the nature of space as a social product is concealed in part by the 'illusion of transparency' which renders space 'luminous' and 'intelligible', 'giving action free rein' (Lefebvre, 1991: 27). It 'goes hand in hand with a view of space as innocent, as free of traps or secret places' (ibid: 28). In this sense space is not seen as socially produced but as unproblematic and neutral and, therefore, knowable and understandable. These ideas about space are bound up with the dominance of the visual as a way of knowing in Western culture, with what Lefebvre terms the 'logic of visualization'. This notion and its associations with a single masculine viewpoint are explained briefly in the Introduction and explored in some detail with reference to dance and architecture in Chapter 10. This connection with the visual means that when space is rendered transparent, as Lefebvre claims, 'everything can be taken in by a single glance from that mental eye which illuminates whatever it contemplates' (ibid: 28). As is pointed out in the Introduction the single masculine viewpoint associated with ideas about perspective is based on the premises of Cartesian dualism that separate out 'body' from 'mind' and 'self' from 'world' and construct a notion of the subject as rational, unified and distanced from the world (see also Chapter 8).

These same premises underlie Lefebvre's concept of transparency and Rose's of the geographer's gaze. She argues 'the claim to see all and therefore know all depends on assuming a vantage point far removed from the embodied world', and 'this transcendent, distanced gaze reinforces the dominant Western masculine subjectivity in all its fear of embodied attachment and all its universal pretensions' (1993a: 70–1). Consequently the illusion of 'transparent space' means that, as Rose suggests, it can be known 'only through a certain masculinity' and that 'like the masculine subjectivity on which it relies, transparent space hides what it depends on for its meaning: an other' (ibid: 71), which Rose sees as 'place'. She suggests 'in this dualistic structure of meaning, masculine knowledge of lucid transparent space is made sense of only in contrast to a notion of "place" as an unknowable' (ibid). In other words 'space' and 'place' constitute a dualism that can be aligned with others such as 'mind'/'body', 'self'/'other' and 'male'/'female'. Characteristics of this 'transparent' space according to one geographer Rose cites, Gould, are its 'infinitude', its 'unboundedness' and the freedom it provides 'to run, to leap, to stretch and reach out without bounds' (ibid: 75). She comments 'these claims of power over space ... suggest to me that [the] space is that of hegemonic masculinity. Only white, heterosexual men usually enjoy such a feeling of spatial freedom. Women know that spaces are not necessarily without constraint; sexual attacks warn them that their bodies are not meant to be in public spaces' (ibid: 75–6). The implications of seeing space from this gendered perspective are explored in the readings of the dances that follow.

Suggestions of travel in the dances

In postmodern dance, travel is not being used as a metaphor in the same sense as it is in postmodern discourse. The dances examined here focus on and create images of, or associated with, 'real' travel, but in the sense that any dance or art work is re-presenting ideas about the 'real world', I would argue that these representations of travel sometimes act as metaphors.

Four dance films associated with travel of different kinds are examined: *Carnets de Traversée, Quais Ouest* (1989, video: 1990), translated as 'Crossing Notebooks (notes or jottings), West Quays', *La Deroute* (1990, video: 1991), 'the Route', *Land-Jäger* (1990), literally translated as 'Land or Country Hunter' and *Cross-Channel*, (1991).³ Their titles suggest travel and they include a range of locations that connote travel such as quays, ports, stations, railtracks, roads, a hotel, open landscapes and beaches. The travel suggested can be seen to fall into three loose categories; migration, exploration and tourism. *Carnets* and *La Deroute* are described as 'dance performances based on the theme of migration' and *Land-Jäger* as 'an imaginary journey' (Éditions à Voir, 1990, 1991). I suggest this journey is about exploration, since it includes activities for three men, such as striding through high undergrowth and sitting and sleeping round a campfire. There is also no evidence of any 'civilization' in the landscape. *Cross-Channel*, which features cycling, camping, staying in a hotel, swimming, sunbathing and partying on the beach, concerns holidaymaking or tourism.

Carnets (24 mins), choreographed by Johanne Charlebois and directed by Harold Vasselin, shows a group of six 'migrants' dressed in European, turn-of-the-century peasant garb, three men in caps, braces, shirtsleeves and trousers, and three women in mid-calf length skirts. The costume together with the use of monochrome suggests a time in the past – the publicity synopsis mentions, 'a hypothetical journey to the New World . . . the promised land'. *Carnets'* title and the quayside location where signs such as 'Acces au Paquebot' 'Access to boat' are in French and English, indicate the migrants' departure from a French port. *Carnets* was in fact filmed in Le Havre. The locations include a departure hall, customs hall, the quayside and harbour wall and water, with shots of docks and cranes in the distance, and the metal ladders and gangplanks of a ship, all suggesting a port. As the film opens, the 'migrants' are seen happily dancing the Charleston in the departure hall, to an old sounding recording of Abe Lyman's California Orchestra playing 'Shake that thing', indicating their imminent, hopeful departure to the 'promised land' of the USA. Prior to the title credits, an enigmatic female figure with feathered turban and trailing tail, a mythical, half-woman, half-bird, surreal creature of flight, is seen ascending a vertical metal runged ladder. She appears repeatedly throughout the film, often in high places looking down on the quayside and the migrants, who seem oblivious of her. The publicity synopsis states, 'A woman appears. . . . Is she real or just the reflection of a desire to know the other side of the ocean?' In between her appearances, the migrants are seen standing in a line in the water up to their ankles facing out to sea, some holding suitcases or boxes; standing, sitting and dancing on the harbour wall; walking, running and dancing in the dock area among warehouses and packing cases; and running and lifting each other up and down the gangplanks of a boat. Pedestrian movements are mixed with more expressive contemporary dance movements of the kind associated with 1980s 'Euro-crash'.[4] *Carnets* is accompanied by natural sounds – the dancers' feet on the harbour wall, ships' sirens and seagulls – occasionally interspersed with Jean-Jacques Palix's music of accordion chords, drum rolls and a persistent metallic sounding beat.

The publicity for **La Deroute** (26 mins), choreographed by Tedi Tafel and directed by Rodrigue Jean who also both perform, states the film is 'about immigrants battling the elements in a cruel and unfamiliar land'. Filmed in monochrome, it opens with a scene of about forty men, women and children standing alone or in small groups spread out across a windy beach. These people by their dress of current day coats, trousers and skirts, look more like contemporary migrants than those in *Carnets*. The film follows a central female character as she wanders along a shoreline, across various tracts of open windswept land; from the beach to mud flats, to grasslands, to sandbanks in water, through a dense wood out into the open to further expanses of flat land and marshes. She is occasionally followed or accompanied by a man and she encounters other travellers along the way. As in *Carnets* the movement material mixes pedestrian postures – walking and running – with more expressive contemporary dance movements. Perhaps the similarity of style is not surprising since the two pieces are featured on the same video, Rodrigue Jean, the filmmaker of *La Deroute*, performs in *Carnets*, Johanne Charlebois, the choreographer of *Carnets*, is credited as Assistant Realisateur on *La Deroute*, and the two works share a dancer, Veronique Favarel. The accompaniment for *La Deroute* is a mixture of natural sounds such as the wind, peoples' distant voices and people walking through water and occasional atmospheric music by Monique Jean when percussion or violins suggest suspense or solitude.

Land-Jäger (13 mins), choreographed and directed by Stefan Schneider, also filmed in monochrome, features three men in dark suits and light shirts, in various outdoor locations; a beach, the sea, a grassy field, a wood, a hill and a maize field. They are seen buried in sand, in the water, walking and running over the hill, crouching in the earth, striding through the maize field and seated and lying around a campfire. The movement material mixes pedestrian activities with episodes of more stylized dance movement. For example, one of the men performs an athletic solo, looking as if he is being continually knocked to the ground in a fight. He repeatedly bounces back in the style of a robotic break dancer. At times the film uses fast freeze frame edits from one shot to another animating the performers' postures. Other surreal elements include a transparent cube that is seen superimposed over a grassy field at the outset, over the campfire, and on the beach at the end. The accompaniment is a mix of natural sounds such as lapping water, and African traditional xylophone and drumming music from Douala, Ethiopia and the Nile.

Cross-Channel (25 mins), choreographed by Lea Anderson and directed by Margaret Williams, follows two groups – one consisting of seven women and the other of five men – from London across the English Channel to Calais in France. The film repeatedly alternates between showing the women and the men. The women travel by train from London's Victoria Station to Dover where they board a cross-channel ferry for Calais. In France they are seen in a hotel, on a beach, in a café/bar with the men, and with them again at a party in a beach hut. The men travel by bike, they are seen cycling across London past recognizable landmarks like Big Ben, through the English countryside to the top of the white cliffs of Dover where they camp. They then appear as workers on the cross-channel ferry, directing cars on and off and cleaning windows and steps, and in Calais on the beach bathing, in the café/bar with the women and at the final beach hut party. The women are dressed smartly throughout, in stylish, white, fifties-style summer dresses and matching hats with black trim and accessories for the journey, in black and white fifties-style swimsuits and sunglasses for the beach, accompanied by hats and feather boas in the café/bar, and in colourful short cocktail dresses with feather trim for the end party. The men are costumed more casually in cycling gear, matelot-shirts and shorts when camping, overalls when working on the cross-channel ferry, old fashioned black bathing costumes on the beach and suits for the party. Each group is costumed identically or similarly making them stand out as groups of performers, rather than blend in with other people in the various locations. Unlike the other three pieces *Cross-Channel* is filmed in colour and in a mixture of urban and rural, peopled and empty spaces. The movement material is pedestrian throughout but it is often repeated and patterned in the choreography so that it looks performed. For example, the men's circular cleaning of the boat windows becomes a repeated rhythmic phrase, as does their dealing of playing cards in a tent when camping. *Cross-Channel* is accompanied by Steve Blake's jazz music – a continuous beat on saxaphone, trombone, bass and drums – over which natural sounds, such as the train and ferry announcements, can be heard.

Gendering of travel, space and subjects in the dance films

Transparent space

The gendering of space and travel that is evident in some postmodern discourse is also apparent in the dance films. *Land-Jäger* and *La Deroute* feature wide open spaces of the kind that Rose suggests can be rendered 'transparent' by the geographer's gaze. In both films figures are seen in the distance, running across windswept, open expanses of grassy fields. The filming and choreography construct the space as infinite and unbounded and, following Rose, viewed from a masculine perspective. There is also a sense in three of the films – *Carnets*, *Land-Jäger* and *La Deroute* – that the space appears innocent and 'transparent' because it is not clearly identifiable as a specific *place*. A generalized, almost abstract sense of space, rather than a specific place is being evoked.

Massey suggests that what is needed to expose the 'transparency' at work is 'a progressive sense of place', 'an understanding of "its character", which can only be constructed by linking that place to places beyond' (1993: 68). Massey is suggesting dismantling the binary of 'space' and 'place'. She continues, 'it would be about the *relationship* between place and space . . . a global sense of the local' (ibid: 68). This is what is presented in *Cross-Channel*. The characters in the dance are clearly situated, they are identified with places beyond the channel crossing and the French coast by markers on their journey, including, for example, London's Victoria Station and Dover's white cliffs. Importantly also the social construction of the space, and the role that gender plays in the construction, are not concealed by the illusion of transparency. Signs of gender self-consciously characterize the places passed through on the journey. The group of women dressed stylishly in their fifties-style white dresses, looking like fashion models from *Vogue* magazine, are seen at Victoria Station, on a train, on a cross-channel ferry and in a hotel. They are clearly gendered traditionally feminine. The construction of their femininity is underlined by repetition of costume, all wear colour co-ordinated variations of the same fifties-style dresses, by inclusion of specifically feminine postures and gestures, such as photo poses with one hand behind the head like a model and adjusting their sunglasses affectatiously and by the filming. When on the train one of the women's faces is filmed in profile, in close-up, looking out of the window, the camera lingers on her made-up face and neatly styled hair. Associations with ideals of femininity featured in adverts are suggested. These women feminize the spaces, in the way that models do on tourist brochures, they make them seem attractive. Their femininity is not hidden or masqueraded as 'natural', it is rather paraded as constructed. By association, the places they inhabit and pass through and the mode of travel they engage in are also gendered glamorously feminine.

Nostalgic space

The feminist art critic Sarah Kent claims that postmodern painting is 'a form of mourning for lost power, lost belief and lost confidence, in which actual significance is replaced by overblown self-importance... It is a masculine artform – a witness to the crumbling of certainty' (1984: 61). In a similar sense the use of travel metaphors in postmodern discourse can also represent mourning, but for different kinds of lost masculine, colonial power. Some theorists have proposed that postmodernism is 'a response by intellectuals to their own discomfiture, their sense of dislodgement from previous authority' (Massey, 1991: 33). Practices of decentering and destabilizing that postmodernism champions, clearly pose threats to established, centered power bases which are traditionally masculine. A mode of thinking which sets up a centre and a periphery in opposition to one another, apart from being modernist, is also imperialist and colonialist (Docherty, 1993). As the postmodern theorist, Steven Connor, suggests, the use of 'metaphorical–topographical terms of space and territory' which include 'centre and margin, inside and outside, position and boundary... can conjure up an oddly antique-seeming map of the world and global political relations, when struggles for power and conquest could be represented in much more reassuringly visible terms.' He continues, 'in their mimicking of this ... vanished territorialization of power relationships, these metaphors also seem to embody a nostalgia for what has been lost with that sort of map of the world' (1989: 227).

Mapping, through its global visualization, positions and controls the 'other' which includes the feminine.[5] This can be compared with the traditional treatment of landscape in painting as 'inanimate' (Oulton quoted in Lee, 1987: 23) derived from the male artist's search for 'beauty' and perfection, and seen in terms of his ability to transform and subjugate 'nature', where 'nature' is traditionally constructed as feminine and 'other' (see also Chapter 4). This parallels the filming and construction of the landscape, nature and performers in *Land-Jäger* and *La Deroute* which render them 'transparent' or innocent. There is also evidence of a 'mourning for lost power' of the sort Kent suggests, since associations with nostalgic notions of territory as available for exploration, possession and colonization, are apparent. An example is the hunting theme of *Land-Jäger*, evident in its title and shots of hands taking live fish out of water, together with other boy scout/Iron John/new man activities of the male performers. They sit round a campfire, sleep out under the stars, and stride purposefully through a field of high maize, pushing it aside as they go. A sense of conquest of nature is emphasized when the camera cuts through the maize, giving the audience parallel experiences of tramping through and over 'nature'. A yearning for lost colonialism is also suggested by the accompaniment of African traditional music which creates an eerie, almost jungle-like atmosphere. Several long shots of a triangular structure in water in *Land-Jäger* could suggest rationality through association with classical geometry. This in turn could be associated with the masculinity of the men in those suits, which dissociate them from the femininity of the landscape. Nature's association with the feminine is also evidenced in *La Deroute* when the central female character dances a long solo in a large marshland puddle. She gradually gets wetter and wetter, until she sinks to her knees in the muddy water, repeatedly arching her back revealing her neck and throat and expressively circling her head and torso in violent whole body gestures. In the filming her circling arms and upper body merge with sprays of muddy water. Her wetness blurs with that of the puddle – woman and nature become one.

Spaces evocative of romantic travel

Travel can be seen as a response to the postmodern crisis of subjectivity when viewed as a metaphor for a search for 'self' or a quest. It has been claimed that tourists 'embody a quest for authenticity' (Urry, 1990: 8). However, as Wolff points out, this metaphor is also gendered in that 'men have . . . an exaggerated investment in a concept *of* a "self" ' (1993: 231). The exploration of *Land-Jäger* has strong associations with a search for 'self'. There are repeated shots of one or two men walking away from camera across a grass field towards a wood, although there is no apparent destination, a sense of purpose is indicated by repetition. Absence of a clear destination and proximity to nature often characterize romantic quests. Romantic ballet narratives such as *La Sylphide* (1832) and *Giselle* (1841) come to mind, where male heroes pursue female spirits of nature in search of an ideal. The feathered female figure, who mysteriously haunts *Carnets*, is like a Romantic ballet spirit in her potential for flight and symbolism of freedom. She is often seen ascending, above the migrants, looking down over them, or into the distance. They do not acknowledge her, she seems to be from another world. As the publicity states, 'Is she real or . . . the reflection of a desire to know the other side of the ocean?'

This 'desire to know the other side of the ocean' is also evoked in *Carnets* by dancing set on the quayside or harbour wall often involving looking out to sea. There is a sense in which postmodern tensions between the global and the local, or space and place, could be said to be suggested by relations that are constructed between the performers and the land and seascapes presented. Distance shots of land or sea and sky show *horizons* which have a particular metaphorical, also often romantic or mythic resonance, in terms of travel, especially when figures are placed looking towards them as they are in *Carnets* and *Cross-Channel*. Such scenes conjure up images from familiar Romantic paintings such as Caspar David Friedrich's 'The Monk by the Sea' (1808–10), where the size of the figure – small, almost minute – in relation to the size of the land or seascape – vast – is powerfully evocative of a certain kind of liminality. This is a romantic, masculine view of the world where nature is seen as 'other', evoked in *Carnets* by several shots of people looking out to sea, one of which significantly ends the film. In *Cross-Channel* performers are also seen looking out to sea, but they are constructed differently. Instead of a single, lone figure gazing romantically towards the horizon, there is a line of equally spaced figures gazing out to sea at sunset. Their performance self-reflexively underlines the image's construction. They act out the cliché prevalent in romantic fiction and film. They are not 'natural'.

The power of the gaze
Rose describes the geographer's gaze as 'penetrating', associated with a masculine viewpoint and with a 'strong claim to knowledge', based on

the illusion of the 'transparency' of space which appears to be unmediated and 'utterly knowable' (1993a: 70). Theories of the 'look' or the 'gaze' and the ways in which looking is gendered have been prevalent in feminist scholarship from John Berger's (1973) statement that 'men look' whilst 'women appear' and connote 'to-be-looked-at-ness', to Laura Mulvey's (1975) theory of the male gaze at work in Hollywood cinema, to Luce Irigaray's statement that 'investment in the look is not privileged in women as in men. More than the other senses, the eye objectifies and masters' (1978: 50).

Associations of objectification, mastery, possession, exploitation and voyeurism with the masculine gaze relate to those travellers who were also explorers and conquerors in colonial times, as well as to tourists today. The use of telephoto lenses in game parks comes to mind. In the sense that tourism is 'a form of entertainment dependent on exploitation' (Chaney, 1994: 79), a notion of 'the tourist gaze' has been suggested (Urry, 1990) and the tourist has been described as a voyeur (Chaney, 1994: 173). The geographer's gendered gaze can be synthesized with Mulvey's masculine gaze since both objectify and rationalize from a masculine perspective; the geographer's gaze objectifies the land and the masculine gaze objectifies woman.

The synthesis of these two gazes is demonstrated in the dance films in the way that space is rendered 'transparent' by the filming and in the way that the women in some of the pieces are also presented as 'transparent', when clearly they are gendered by certain filmic and choreographic devices. As Mulvey claims both the camera and the other (male) performers position the viewer to see the female performer(s) through the masculine gaze.

The men in *Carnets* are seen actively moving through space; climbing steps, walking along the quayside, jumping on the quay wall. The women are often seen looking, out across the water and into the distance, nearly always stationary, often in close-up, their faces well-lit with wind-blown hair. When the women dance it is often in a sensitive, interior way. In *La Deroute* there are similar close-ups of women's faces and the long solo of the woman in water which arrests the narrative in just the way Mulvey describes for mainstream female Hollywood film stars. Unlike the characters in *Lament of the Empress* described in Chapter 1, the filming of this woman and her performance construct her as an object of the masculine gaze. She is filmed in close-up and from overhead and her performance consists of expansive, expressive, typically feminine, modern dance gestures. Both the woman and the land, in the sense of mud and water, are constructed as 'transparent' and objectified by the gaze.

Cross-Channel also presents women as passive and to-be-looked-at – sitting in the train, leaning over the side of the ferry, sunbathing on the beach – and men as active – cycling, directing traffic onto the boat, cleaning the ferry windows and swimming. The choreography in *Cross-Channel* also emphasizes gender differences. The women are given mainly whole body, fluid movements, as in a hotel corridor when they emerge from doorways, spin across the hallway and disappear into the rooms opposite. Whereas the choreography for the men consists mainly of rigid movements of isolated body parts, such as when they swat flies outside their tents on the clifftops, or when they gingerly dip their toes in the sea, wavering as they paddle. The choreography, particularly when combined with the costumes, results in the women looking suave, sophisticated and in control, and the men looking clumsy, awkward and silly. The women appear as competent 'natural' travellers, whereas the men look uncomfortable in this role, reversing, and hence questioning, the dominant viewpoint. Through various structural devices such as editing and repetition, the passivity of the women and the activity of the men in *Cross-Channel* are emphasized, foregrounded and played with and thus exposed and critiqued through ridicule. Gender is revealed as constructed rather than hidden by the illusion of transparency as it is in *Carnets* and *La Deroute*.

Reappropriating travel metaphors for new subjectivities

Wolff concludes her article on travel metaphors by proposing the *reappropriation* (her emphasis) rather than the avoidance of such metaphors. A reappropriation is, in her words, 'a good postmodern practice which both exposes the implicit meanings in play, and produces the possibility of subverting those meanings by thinking against the grain' (1993: 236).

The American feminist Donna Haraway's essay *The Promises of Monsters* (1992) can be seen as an example of Wolff's proposed reappropriation of travel metaphors since she exposes and subverts the meanings in play and thinks 'against the grain'. Haraway states that the essay will be a 'mapping exercise and travelogue through mindscapes and landscapes of what may count as nature in certain local/global struggles'. The purpose of the exercise is to write theory that will 'produce not effects of distance, but effects of connection, of embodiment and of responsibility for an imagined elsewhere' (1992: 295). One result of this is to see 'nature' as constructed. Haraway proposes 'we must find another relationship to nature besides reification and possession . . . in this essay's journey toward elsewhere, I have promised to trope nature through a relentless 'artifactualism' (ibid: 296). By 'artifactualism' she means that nature is constructed like an artifact. It is *made* as fiction and fact.

A travel metaphor suggested in Haraway's 'journey to elsewhere' is 'networking' resulting from 'affinities'. Haraway says, 'I prefer a network ideological image, suggesting the profusion of spaces and identities and the permeability of boundaries in the personal body and in the body politic' (1990: 212). This proposal of a network image addresses, for her, the fluid, multi-layered and fragmentary nature of postmodern subjects. She is proposing a connected notion of embodied subjectivity associated with the feminine. This is very different from the subjectivity constructed in some of the dance films where seeing space and travel as 'innocent' and 'transparent' hides the gendering at work. When space is seen as distanced and unknown as in *La Deroute*, *Carnets* and *Land-Jäger*, the subjectivity suggested is detached, it connotes the power and exploitation of travel associated with exploration, colonialism and tourism. From this masculine perspective of the geographer's gaze, space is objectified as other and seen as feminine. The viewpoint is distanced and detached from the world with associations of power over it. Haraway's concept of a network ideological image in contrast privileges seeing from more than one perspective over the single penetrative masculine gaze. A 'profusion of spaces and identities and the permeability of boundaries' are posited.

This view of space is evident in *Cross-Channel* where connections and networking rather than distance are suggested by the rapid editing cuts from the group of females to the group of males, and by the many spatial conjunctions of the two groups, which increase in proximity as the film progresses. The male cyclists ride under a railway bridge over which the train, which presumably carries the women, is crossing. The men work on the cross-channel ferry on which the women travel. Finally, the two groups share the same beach and café/bar, and attend the same party. *Cross-Channel* also suggests the 'permeablility of boundaries in the personal body and the body politic' that Haraway mentions. It does this by exposing the constructed nature of gender showing that those personal body boundaries are not fixed but permeable, and by revealing the power invested in such constructions, because they are associated with certain kinds of activity or passivity.

Conclusion

In the light of these theories, it is evident that *Carnets*, *La Deroute* and *Land-Jäger*, reinforce androcentric worldviews. They do this by presenting 'transparent' views of space and of women from a masculine perspective, rather than revealing situated pictures of places. They also privilege a penetrative, masculine gaze which objectifies. The results reinforce constructions of landscapes as inanimate and available for possession, of nature as reified, passive and feminine, and of journeys and travel as active masculine pursuits in search of an ideal self. By doing this, these texts conceal the politics and power differentials of different kinds of travel, space and subjects. The powerlessness of the migrants of *Carnets* and *La Deroute* is eclipsed by the 'beauty' of the filming and choreography. These constructions of travel and space perpetuate patriarchal notions of subjectivity that is fixed and distanced from the world and associated with power over it. In contrast *Cross-Channel* presents clearly identifiable places rather than anonymous spaces, it *situates* the performers in them, and, through self-reflexive filming and choreography, shows 'otherness' and gender to be constructed rather than 'natural'. By working against the grain in this way, *Cross-Channel* re-appropriates the travel metaphors and subverts the more masculine and androcentric readings of travel, space and subjects. In the light of Haraway's theories, the focuses on situated places over transparent spaces can be seen to privilege notions of networks and affinities with the spaces. These ideas suggest possibilities for new ways of thinking a less distanced and more permeable, connected, embodied subjectivity that I believe can sometimes be imagined in postmodern works like *Cross-Channel*.

The next chapter focuses on the particular ways in which dancing bodies and the spaces of their performance can mutually construct each other, and in the process transform the spaces and subjects involved.

3
Transforming City Spaces and Subjects

Introduction

During the nineties the city as a paradigm example of postmodern construction became the focus of considerable attention (see for example Clarke, 1997; Lefebvre, 1996; Pile, 1996; Watson and Gibson, 1995). The use of cities as settings for postmodern dance, in live site specific performances and in dance films and videos, was prevalent in the late eighties and early nineties.[1] This chapter examines one European and one British postmodern dance video; *Muurwerk* (1987), choreographed and performed by Roxanne Huilmand and directed by Wolfgang Kolb; and *Step in Time Girls* (1988) choreographed by Yolande Snaith and directed by Terry Braun. Both use the city as a setting for female solos. Following the French post-Marxist theorist Henri Lefebvre, who proposes, 'each living body is space and has space: it produces itself in space and it also produces that space' (1991: 170), I examine how the city spaces and the dancing bodies mutually construct each other and the role gender plays. The city is a particular kind of constructed space and its construction is inextricably bound up with constructions of subjectivity. 'To the extent that the inhabitant of the (post)modern city is no longer a subject apart from his or her performances, the border between self and city has become fluid' (Patton in Watson and Gibson, 1995: 117–18). In the analysis of *Muurwerk* and *Step in Time Girls* particular attention is paid to the ways in which the dancers' interactions with the urban environments, with the fluid borders between themselves and the city, contribute to the construction of spaces and subjects.

Earlier instances of dances located in city spaces evident in avant-garde performances of the sixties provide key precedents for this work.

The conceptual focuses of sixties artists on the avant garde use of site specific performance spaces which stretched audience perception, on a particular urban sensibility and on blurring boundaries, such as inside/outside, private/public and art/everyday life, paved the way for what was to follow. Examples of this work by American choreographers, Lucinda Childs, Meredith Monk, Twyla Tharp and Trisha Brown provide an historical context for the later works.

Sixties precedents

In New York in the sixties dances were often set in city spaces such as streets, museums, lofts and parking lots. The significance of using such spaces had been proposed by artists such as Allan Kaprow and Claes Oldenburg when they began to stage 'Happenings'. Kaprow, who wanted to blur the boundaries between 'art' and 'life', wrote, 'we must become preoccupied with . . . the space and objects of our everyday life . . . our bodies, clothes, rooms. . . . Forty-Second Street . . . happenings and events, found in garbage cans, police files, hotel lobbies; seen in store windows and on the streets' (in *Artnews* in 1958 quoted in Crow, 1996: 33). Oldenburg used the Judson Memorial Church gallery space in Washington Square, New York, which he and friends had established in 1959, for his installation *The Street* in 1960. It consisted of 'crudely fashioned props and figures . . . intended to evoke the . . . life of the poor neighbourhoods . . . where he . . . lived and worked' (Crow, 1996: 34). In *The Street* he staged a Happening entitled *Snapshots from the City* – 32 tableaux each appearing briefly before being blacked out. Oldenburg stated he was coming to grips 'with the landscape of the city, with the dirt of the city, and the accidental possibilities of the city' (ibid: 34). A similar urban sensibility informed certain dance performances in sixties New York, and infuses the recent fascination with the city evident in the two dance videos examined later.

In 1964 Lucinda Childs, one of the dancers from the Judson Dance Theater, named because of their performances in the eponymous church, created *Street Dance*. She and another dancer performed in the street four or five floors below the Cunningham dancer, Judith Dunn's studio, where a tape instructed the spectators to watch from the window. Childs said, 'The dance was entirely based on its found surroundings . . . we were engaged in pointing out . . . details and/or irregularities in the façades of the buildings: lettering and labels, the . . . displays in the store fronts. . . . While the spectators were not able to see in . . . detail . . . what it was we were pointing to, they could hear the information on a tape' (in Livet, 1978: 61–3).[2] Childs mentions the dance blending 'in with the other activity...in the street' citing an incident when a pedestrian asked her a question and she stopped what she was doing to answer him. She continues, 'I liked that, the fact that the dance could fit into this self-contained setting where everybody, including me, was going about their business' (ibid). This engagement with the urban environment blurred the boundaries between performance and everyday life, coinciding with these artists' conceptual investigations of the limits of dance and performance. They were examining what it meant to dance and perform, where such performances might take place and the basic elements of space and time in which performances existed.

The spatial configuration of *Street Dance* stretched the audience's perception. This was an important consideration for Childs. She said, 'the spectator was...called upon to envision, in an imagined sort of way...information that...existed beyond the range of his... perception' (ibid: 63). Childs admits that this experiment with perception proved important for her later work (in Kreemer, 1987: 97). Viewing dancers from a distance through glass windows and walls resurfaces in later postmodern dances filmed in cities, such as *Lament of the Empress* (1989) cited in Chapter 1, *Step in Time Girls* examined in this chapter, and *Duets with Automobiles* (1993) and *Rosas Danst Rosas* (1997) examined in Chapters 6 and 10 respectively. When dancers are framed and filmed through glass, exploiting the fluid borders between self and city, a mutual construction of the dancing bodies and the spaces which they inhabit can result. Plays with perception of this kind affect the way in which performers are viewed as subjects or objects and have an important bearing on constructions of subjectivity. In this sense the avant-garde nature of works like *Street Dance* opened up possibilities for a consideration of the more overtly political concerns of feminists which surfaced in the seventies, and are key to the gendered constructions of subjectivity examined in *Step in Time Girls* and *Muurwerk*.

Another work where performers were seen through windows was Meredith Monk's *Blueprint* (1967), first performed in Woodstock, and later in the Judson Gallery and Monk's loft in New York. In the Woodstock performance spectators viewed live and filmed events through the windows of a building from outside, before moving inside to see effigies of the figures who had just performed, through the windows outside. Banes comments, 'Throughout, there was a confounding of inside and outside' (1980: 161). She sees this as the basis for many metaphors in Monk's work, extending the blurring of boundaries to 'the borders between private and public... the body and the universe [and] the individual and community' (ibid). As Harrison and Wood comment, the site specific artworks of the sixties 'generated a phenomenologically informed focus upon the conditions of encounters with artworks: ... the embodied perception of physical objects and events' (1992: 799). Phenomenologically informed embodied perception is one of the bases for the Australian feminist critic Elizabeth Grosz's theories of mutual definition of bodies and spaces in general, and of bodies and cities in particular, which I use in this chapter. Grosz uses the term 'interface' to describe a 'two-way linkage' which she suggests exists between bodies and cities (1995: 108). This can be seen to be consonant with the blurring of boundaries between events and audiences in the sixties performances, characterized as 'embodied perception' by Harrison and Wood, and with the dissolveable divisions Banes sees in Monk's work. The importance of particular sites seen from different perspectives together with the recognition of the fluidity of borders, and the significance of the spaces in between, are apparent in much of Monk's later work, as the examination of her *Ellis Island* (1981) in Chapter 7 illustrates. The blurring of actual public and private spaces, evident in Monk's performances in her own loft, is also a feature of *Step in Time Girls* and other current site specific work (see Chapter 5). The use in *Blueprint* of particular parts of buildings such as windows and the roof or loft – in the Woodstock version a man stood on the roof and in the Judson performance there was a man in the church loft (McDonagh, 1990: 114) – while playing with the spectators' perception, also focuses their attention on the frames and façades of the urban environment. People can seem confined and constrained when seen through window frames or against façades,[3] highlighting features of urban existence foregrounded by similar means in *Step in Time Girls* and *Muurwerk*.

Specific sites for Monk are important, she feels 'each piece should grow partially out of the site in which it is performed' (McDonagh, 1990: 117) (see also Chapter 7). This is apparent in her use of the Wooster Parking Lot in *Vessel* (1971), where among many other things, she lit up the portico of the church across the street. Commenting on this, Monk said, 'I'm working like a filmmaker', 'it's like expanding the environment until you're aware of more and more. ... When you think you've got to the limits of the parking lot ... your eye moves across the street' (in Cohen, 1992: 216). Monk, in common with other sixties New York artists, was investigating the limits of performance by expanding her audience's perception. The parallels with filmmaking she recognizes, where the artist can play with audiences' viewpoints in a range of ways, become commonplace in dance films of the eighties and nineties, as the use of the camera in the two works analyzed in this chapter demonstrates. The specificity of the Wooster parking lot is also important for Monk. She describes it in detail, 'On one side... is the Canal Lumber Company and on the other side... is a church, St. Alphonsus... straight ahead of the audience is a candy factory, an old building with a faded sign on it and a few straggly trees at its base' (ibid: 215). This description is reminiscent of Childs' *Street Dance* score. Both choreographers clearly had an eye for the details of their city. Monk underlined the importance of this particular city space when she said, 'I was trying to find an outdoor space that had a specific New York ambience... one level of *Vessel* has to do with people opening up their eyes to New York' (ibid: 215). This concern with an urban sensibility is also evident in Monk's construction of the space – as well as lighting it, she installed among other things furniture, eighty performers sitting round campfires and a motorcycle cavalcade. She created a particular place using the parking lot's specific ambience. A similar kind of attention to the detail of city spaces is evident in the work of the later postmodern choreographers examined here.

Twyla Tharp's *Dancing in the Streets of London, and Paris, Continued in Stockholm and Sometimes Madrid* (1969) was first performed on two levels and the connecting stairwell of the Wadsworth Atheneum Museum in Hartford, Connecticut, and later in both the Lincoln Center Performing Arts Library and the Metropolitan Museum of Art, New York. Tharp said 'I designed *Dancing in the Streets* to break down every conceivable wall put up to separate life from art' (1992: 122–5). There had been 'activist interventions within museums and galleries' in New York throughout the sixties as these city spaces were seen as 'outposts of established power' (Crow, 1996: 11). Tharp's intention to blur the boundaries between art and life was consonant with the philosophy behind these actions. Her choreography consisted of 'workaday tasks such as switching clothing, or reading while dancing' (Mazo, 1977: 286). The use of everyday pedestrian movement, which was avant-garde at the time, became a feature of much later postmodern work, such as *Step in Time Girls*, and it can be seen as a source for movement in *Muurwerk*. The specificities of the different museums as performance locales were also clearly exploited, as Tharp indicates, 'we opened by warming up directly in front of the museum entrance so the audience had to walk over and around us to get in . . . the dances occurred in . . . galleries, elevator shafts, broom closets, hallways' (Tharp, 1992: 125). In some ways the performers invaded the museum space reclaiming it for their own purposes. This action, probably inspired by earlier sixties occupations of public art spaces, motivated by a disenchantment with these 'outposts of established power', is paralleled by the metaphorical reclamation of potentially constraining and alienating city spaces by the women performers in *Step in Time Girls* and *Muurwerk*. Tharp made several pieces at this time that used different city spaces; from museums in Stuttgart (*One Two Three*, 1967) and Amsterdam (*Jam*, 1967), to New York's Central Park for *Medley* (1969), and the roof of a school in Brooklyn for her dance film *Stride* (1965).

Trisha Brown famously used roof tops for her *Roof Piece* (1971). This was one of a series of equipment pieces including *Man Walking Down the Side of the Building* (1969) and *Walking on the Wall* (1971), which involved dancers in harnesses doing exactly as the titles describe. These pieces played with the audience's sense of perspective (see also Chapters 8 and 10) and inevitably, because of the tasks set by Brown, the particular city spaces used became part of the choreography. There is a sense in which these urban surfaces were transformed by these performances into challenging edifices for testing the limits of the body's capabilities. Similar transformations of city spaces occur in the later postmodern choreography analyzed in this chapter. *Roof Piece* possibly also inspired part of *Freefall* (1988), a dance video set in London choreographed by Gabi Agis, which has dancers performing on London rooftops spaced out in just the way Brown's dancers were (see Briginshaw, 1997). In both pieces the audiences' perceptions are stretched, boundaries between city space and performers are blurred as the dancers in the distance merge with the city skyline.

The avant-garde use of site specific city spaces by choreographers and other artists in sixties New York stretched audiences' perceptions and blurred boundaries introducing ideas of embodied perception. These ideas made way for an enlightened engagement with mutual constructions of spaces and subjects which, drawing on the achievements of seventies feminists, acknowledges the role that gender plays. The ways in which city spaces and dancing bodies are transformed by this mutual construction, which is recognized as gendered, are examined in the analysis of *Muurwerk* and *Step in Time Girls* which follows.

Muurwerk and *Step in Time Girls*

Muurwerk's (1987–27 mins) title, translated, means 'brickwork' or literally 'wall work'. There is certainly a sense in which Roxanne Huilmand, the solo dancer, is 'working' with the walls of the alley in the old centre of Brussels in which *Muurwerk* is set (see Plate 1). Dressed in a sleeveless, low backed, short skirted dress and boots, Huilmand opens the piece by walking down the alley away from camera, dragging the palm of her hand along the walls. The only sound initially is that of her footsteps echoing in the empty space. What initially appear to be everyday pedestrian movements such as walking and leaning on a wall transform into more obvious contemporary dance as Huilmand begins to rhythmically roll and spin her body along the walls as well as running and kicking against them. These movements are all incessantly repeated and after about 15 minutes they are accompanied by the minimalist violin music of Walter Hus. This 'Euro-crash' style of dancing, is not surprising since Roxanne Huilmand is a former member of Anne Teresa De Keersmaeker's company *Rosas* (see Chapter 10) and De Keersmaeker was one of the choreographers who established Euro-crash (see Chapter 2, note 4). Wolfgang Kolb, who directed *Muurwerk*, has also directed several films and videos for Anne Teresa De Keersmaeker and *Rosas*.

In *Step in Time Girls* (1988–24 mins) three female dancers portraying women from different periods – 'contemporary', 'wartime' and 'Victorian' – are filmed for the most part on their own, inside and outside the same block of low income London flats. Very occasionally they are seen with a male partner, a small child or a baby, but never with each other. While outside the flat, their movements are pedestrian, walking, running and going up and down stairs or steps. When inside, however, they rhythmically repeat and enlarge movements such as rolling along walls and windowsills, rocking in chairs and opening and closing drawers and cupboards. The 'contemporary' woman, danced by Yolande Snaith, becomes quite athletic at times (see Plate 2), running and jumping up on a table in an economically efficient manner. She propels her upper body through a serving hatch, and remaining contained, hangs down from it, and uses momentum to swing along the wall below. Snaith is known for her gymnastic, weighty, 'release-based' style of dancing, characteristic of British 'new dance', developed in the seventies and eighties, particularly at Dartington College of Arts, where Snaith studied under the tutorship of Americans Mary Fulkerson and Steve Paxton. A former member of the Judson Dance Theater, Paxton is noted for his promotion of the athletic duet form known as 'contact improvization'. The editing of the film continually shifts scene from inside to outside, just as some of Monk's and others' performances did in the sixties, and from one woman to another. The women are seen performing similar but not identical movements in the same spaces.

Constructed spaces

Spaces, through their construction, are invested with power. The postmodern city, like the postmodern subject, is fractured and fragmented, it is falling apart and full of contradictions. It is utopian and dystopian, attracting and alienating. It is constructed as labyrinthine, free-flowing and uncontrollable, but also as containing and trapping. The effects of this for some women can be negative; they can be isolated, excluded and constrained in these mainly public spaces historically constructed by men for men.

The two dance video texts discussed here provide examples of different kinds of constructed city spaces invested with power. In *Muurwerk* the deserted alley in the centre of old Brussels can be seen as alienating and trapping, the sort of inner city space that women are warned about and expected to avoid. This is the kind of city space that connotes male freedom associated with the nineteenth-century notion of the *flâneur*. In *Step in Time Girls* the rooms of the London flat and its environs can be seen as unsympathetic urban spaces which isolate, contain and constrain the women that inhabit them. They appear to be a legacy from the 'culture of separate spheres' that the feminist cultural theorist Janet Wolff argues grew up in the nineteenth-century but still persists to a certain extent today. In this culture male and female are closely aligned with public and private spheres respectively giving rise to 'the domestic ideology of the home as haven and women as identified with this private sphere' (1988: 119).

In the dance video texts, the choreographers and filmmakers use discourses which can either reveal or hide ways in which the bodies and spaces are constructed and invested with power. I explore how these texts reveal in a problematic way the possibilities for reclaiming ways of inhabiting the city, which turn anonymous spaces into situated places, by negotiating their negative and alienating characteristics for women. In the process I suggest these choreographers and filmmakers demonstrate possibilities for affirmation and empowerment through defining new relationships with city spaces.

Mutual definition of bodies and cities

In each of these dance video texts the dancers' bodies interact with the city in different ways. These interactions, which involve the physical contact of the dancers' bodies and the fabric of the city, are explored focusing on the mutual construction and definition of bodies and cities. Elizabeth Grosz's notion of the body as a surface of inscription, which she derives from Nietzsche, Foucault and Deleuze, can inform an analysis of these texts in the sense that the power invested in the city spaces inscribes movement patterns on the women's bodies in different ways. Grosz is 'concerned with the processes by which the subject is marked . . . or constructed by the various regimes of institutional, discursive and non-discursive power' (1995: 33). These values and regimes of power, determined by the ways in which the world is seen and conceived, include strands of this conception and construction such as gender. In this inscriptive process 'the body's boundaries and zones are constituted . . . through linkages with other surfaces and planes' (ibid: 34). In these two dance videos the boundaries and zones of the performers' bodies are partly constituted through their linkages with the surfaces and planes of the city. In this sense bodies and cities are 'mutually defining' (ibid: 108).

I am arguing that this process of mutual definition can inform readings of these dances and that through it, the dancing bodies become subjects and the spaces of the city settings become places. This is because exceptionally in the choreography and filming of the dance videos the women go beyond the confined urban spaces. Consequently the mutual definition which occurs can affirm identities for the women contrary to what's expected of them in these city spaces.

Muurwerk

The choreography of *Muurwerk* consists largely of repetitive phrases of rolling, spinning along and jumping up and kicking the walls of the deserted Brussels alley, which appears drab and lonely in this black and white film. There are very few moments when one or other part of Roxanne Huilmand's body is *not* in contact with the alley's surfaces. The rhythmic phrases of rolling and spinning along the walls of the alley are interspersed with Huilmand, rolling across the alley floor into the gutter defiantly revealing her substantial white knickers and with her circling round on the ground from sitting to rolling onto hands and feet. The tactile nature of these interactions is emphasized in three ways: through camera close-ups; choreographically, through persistent repetition and dynamic contrasts of speed and body tension; and aurally, in the soundtrack where the brushing and scraping of body on walls and ground is heard.

As Huilmand repeatedly leans on, pushes off, slides, drags and brushes her body up against the stone sides and pavement of the alley, her relationship with it seems to become increasingly affectionate. The camera, by closing in on her movements, allows her to perform more intimate movements designed to be seen close up, while also emphasizing the intimacy of the space, as it caresses the alley walls and floor. The window ledges, cracks, nooks and crannies become known as they are repeatedly traversed by Huilmand's body. The filming and Huilmand's performance construct and play with 'femininity' in various ways. Huilmand veers between being seductively sensual – when the camera lingers on a close-up of her face as she brushes her hair behind her ear – and aggressively violent as she repeatedly runs, jumps up and kicks against the walls. She teases and titillates through the dynamics and pace of the choreography by very slowly repeating and building up patterns of rolling or spinning along the walls only to abruptly interrupt and change the action before a climax is reached. In this context I suggest that as Grosz argues: 'the practices of femininity can readily function... as modes of guerrilla subversion of patriarchal codes' (1994b: 144). The brazen confidence of Huilmand's apparently fearless performance of a limited but thoroughly known and worked through vocabulary, etched with a subtle playfulness, enables her to transform the normally alien space of an inner city alley into, for her, an intimate place of play and fantasy. Whereas the sixties artists raised audiences' awareness of urban environments and stretched their perception through avant garde uses of space, eighties and nineties' choreographers, such as Huilmand, challenge audiences to rethink traditional gendered relations between female subjects and city spaces.

The choreography and filming of Huilmand's performance make it seem rebellious, and new meanings are created for this body in this city space as they mutually define each other. 'The city... divides cultural life into private and public domains' (Grosz, 1995: 109). A public alleyway such as this is considered a threatening space for women. Following Wolff's identification of the 'culture of separate spheres', it is likely to keep women in their place, in the private, domestic, supposedly safe space of the home. The possibly expected reading of the situation portrayed in *Muurwerk*; a young woman, dressed in a sleeveless, low backed, short skirted dress, alone in a deserted alley, often literally pinned to the walls or floor, is that this woman is likely to be courting danger, that she could be a prostitute. This reading is evidence of the social rules and expectations which are part of the institutional power of the city. As Grosz indicates, 'the city's form and structure provides the context in which social rules and expectations are internalized and habituated' (1995: 109). The form and structure of the alleyway provide a context for expecting women to be in danger and possibly to be prostitutes if they are in such a place. However, this reading is averted by the choreography. Rather than appear vulnerable as might be expected, Huilmand, because of her defiant performance, seems to be making a more affirmative statement about her ability to confidently occupy and inhabit the space of the alley. Her bodily brushes with the fabric of the alley are unremittingly confrontational. Her performance, combined with the choreography and filming, overcomes the expectations associated with her gender in such a city space. She transforms it into a nonthreatening place of surfaces and textures for rolling and spinning along in which she seems at home.

The insistent repetition of what appear to be sometimes angry, sometimes mesmeric, caresses and collisions with concrete in *Muurwerk* flies in the face of any suggestion of vulnerability, which looks instead as if it is being worked out of the dancer's system cathartically. When Huilmand rolls into the gutter and her skirt flies up revealing her knickers, she seems to be defiantly saying 'so what!' As Grosz claims, 'the form, structure and norms of the city seep into and affect all the other elements that go into the constitution of corporeality. It affects . . . the subject's understanding of and alignment with space . . . comportment and orientations. It also affects the subject's forms of corporeal exertion' (1995: 108). Despite the city's impact on corporeality and its power to inscribe a sense of vulnerability on a young woman like Huilmand, as Grosz argues, there is the possibility of a reciprocal reinscription. She suggests, 'the body (as cultural product) transforms, reinscribes the urban landscape according to its changing . . . needs, extending the limits of the city' (1995: 109). In her dancing Huilmand seems to be doing just this, to be at least partly creating her own labyrinth in the alley, discovering new movement and identity possibilities and the potential of new relationships with the city and its spaces. She appears to be reclaiming the space for herself, but differently from the sixties artists such as Tharp and her dancers. Huilmand and Kolb, the filmmaker, are working with discourses of the body and film developed since the sixties, which reveal ways in which Huilmand's gendered body and the space of the alley are constructed and invested with power. They reveal possibilities for reclaiming a way of inhabiting the city, which are through reinscription turning this anonymous alley into a familiar, situated place. The choreography and filming of Huilmand's performance have negotiated the negative and alienating characteristics for women and 'others' of this city space.

Step in Time Girls

First impressions of *Step in Time Girls* suggest that the three women, albeit from different periods, are all contained and trapped by the city spaces they inhabit. They appear subjected by them and made powerless by being forced to repeat movement vocabularies that suggest frustration with their environment and situations. This is apparent when the women repeatedly rock back and forth in chairs, roll along walls and windowsills and stride back and forth along table and windowsill. Their containment, alienation and isolation are emphasized because they are often seen in relation to the windows, their link with the outside world. They look out of them, pace up and down in front of them and dance on the windowsill. They are also viewed from the outside; as the camera recedes, minute pictures of private lives are glimpsed, framed by the numerous identical cell-like windows of the block of flats. The space of the flat is coded 'domestic' by aspects of the choreography; the dogged repetition of tasks that the women perform, such as opening cupboards and drawers, returning a baby's spoon to its highchair tray and running up and down stairs. The choreography abstracts and emphasizes the movements through development, exaggeration and, particularly, through repetition. Different parts of the body, such as shoulders, backs and feet contact the furniture and walls and repetitive rhythms are often built up. The movement patterns these women perform are those which the city has inscribed on their bodies through the power invested in the city spaces that they inhabit. The choreography and filming of *Step in Time Girls*, by revealing ways in which urban spaces contain and trap, gender them traditionally feminine[4] and render the inhabitants, also gendered feminine, to a certain extent powerless.

This is not the whole story however. Through the repetition of these movement patterns the women appear to go beyond the limitations of the city spaces and make them their own. For example, just over halfway through the piece the 'contemporary' woman rolls along the floor of the flat and up to sit on a chair and lean back briefly, she then jumps to sit on a table where spinning round on her bottom she swings her feet up to stand on the table against the wall, from where she looks down at the space she has just traversed. The whole phrase is then reversed and repeated twice. With the reversal and repetition of the choreography a rhythm and phrase are established which appear to be performed and enjoyed for their own sake. The sheer energy and momentum of the dancing takes over. Jumping onto the table, spinning round on it and jumping off again are actions that, because of their audacity in the context of the domestic space in which they are performed, can be seen to be empowering and affirming. The dancer is reclaiming and taking over this urban space for herself. It has become her place.

There are several other examples of choreography where the women appear to stretch and in some cases exceed the constraints of the space. All three women throw their jackets across the space of the flat at one point: the 'contemporary' woman lobs hers up against a window, picks it up and repeats the gesture; three times the 'wartime' woman hurls her jacket at a wall and then 'catches' it on her back as it slides down the wall; the 'Victorian' woman flings her jacket across the room onto a chair. Straight after this the 'wartime' woman spins along a windowsill, jumps up to sit on it and then comes down and, while still holding the upright window support, jumps her hips up high in the air using her other hand to push up from the windowsill. This is repeated. There is a sense in which she is exceeding the limits of the space by her actions, emphasized by the low angled shot which shows her filling the screen. Immediately afterwards the 'Victorian' woman is seen silhouetted against an open window rocking back slowly in her chair balancing on the back legs only, until she tips a little further back and the chair leans against the wall, her image also fills the screen. She then returns forward to repeat the movement – in all it is repeated fourteen times. The last eleven times she raises her arms as her body tips back adding to the momentum. The camera focuses on her feet pushing off the floor and then dangling in mid air as she holds the balance, and on her hands as they are held suspended above her. The movement is slow, it becomes increasingly mesmeric with repetition.

At these moments it seems as if the women have appropriated the space of the flat for their own personal gymnastic or exercise routines. They are enjoying rocking, stretching, balancing and jumping up, using the walls, floor and furniture as apparatus to assist them. This seems particularly the case when much earlier in the video the 'contemporary' woman bursts through the serving hatch, head first, stretching her arms to the side horizontally, holding a gymnast's pose. She balances on her stomach briefly and then drops her upper body down to repeatedly swing it along the wall below. In the next shot she is balancing on her thighs bracing herself with her feet on the upper lintel of the hatch still swinging her torso from side to side along the wall below. She then rocks her upper body away from and towards the wall several times before lowering herself through the hatch down onto the floor in a somersault. These acrobatic and athletic feats go way beyond the norm for women in a domestic interior. This city space is transformed from a containing unsympathetic environment and reinscribed as an imaginative playground where the women revel in the experience of their bodies contacting and rebounding off walls and furniture. This choreography uses the surfaces of the London flat to challenge the capabilities of the dancers in a similar way to Trisha Brown's equipment pieces. However, where Brown was concerned with challenging audiences' perceptions of what and how the human body could perform, Snaith and her dancers in their enjoyment of physicality in this urban environment, challenge viewers to rethink their expectations of feminine behaviour in such spaces. The fact that the three women from different periods perform similar movements in the same block of flats also suggests linkages and connections through time that add to the dancers' construction as subjects. The urban spaces of the London flat and its environs connect the three women as subjects in the city. In these ways *Step in Time Girls* reveals through its choreography and filming possibilities for reclaiming city spaces as places for women. The bodies and the spaces mutually define each other to suggest new meanings.

Conclusion

By placing dances in city settings, choreographers and filmmakers open up possibilities for exploring the ways in which cities and bodies can mutually define and construct each other. It would be possible to use the city, with its multiplicity of signs and images, as an exciting arena for filmed dance performances, without taking account of the ways in which city settings and spaces are inevitably invested with power; simply using the city as a backdrop in an uninformed manner that takes it for granted. Some choreographers and filmmakers have done this (see Briginshaw, 1997). However, the choreographers of the sixties cited and the choreographers and filmmakers of the later dance video texts examined, have gone beyond simply using the city as a backdrop. Through their choice of settings, their choreography and performances and through the filming, they have revealed some of the ways in which bodies produce themselves in space and in turn produce space (Lefebvre, 1991). The possibilities for this kind of transformation, opened up by the avant-garde sixties site specific urban performances, are evident in the two dance videos examined.

In *Muurwerk* the deserted, drab Brussels alley is gradually transformed through the solo female dancer's almost hypnotic repetitive phrases of rolling, spinning, sliding and kicking. This city space, that connotes male freedom associated with the nineteenth-century notion of the *flâneur*, is reinvested with a distinctly feminine kind of potency, as, through her dancing, Huilmand weaves her own thread of power and takes on the mantle of a contemporary *flâneuse*. The anonymous space becomes an imaginative situated place of play – of musings and teasings, suggesting new meanings.

In *Step in Time Girls* the rooms of the London flat and its environs which contain and confine its three female inhabitants from different eras, through the choreography and filming reveal the potential for change. The women's performances, by first reiterating the boundaries of the spaces they inhabit, set up possibilities for transcending these limits. It soon becomes apparent, through the choreography and filming, that these women have the potential to go beyond the constraints of their environment, just as Brown's performers did in the sixties. Despite the expected internalization of rules and habits which the city inscribes, through the repetition and development of choreographic phrases, which increasingly visibly become enjoyed for their own sake, Snaith and her co-performers, by revelling in the sheer physicality of their movement experience, each in their own way, reclaim and reinscribe these city spaces for themselves.

The examination of these dance video texts and the earlier precedents from the sixties, has shown that because both bodies and cities can be seen as constructed, there is always potential for change, and that when bodies and cities coexist and interact, there is also potential for their mutual construction and definition. As the two dance videos have shown 'there is nothing intrinsic about the city that makes it alienating or unnatural' (Grosz, 1995: 109). The ways in which these dance video texts and the earlier sixties work have facilitated the construction of bodies and city spaces as subjects and places, have indicated the potential for new meanings. The site specific work of the sixties began to erode the conventional boundaries between performers and audiences such that the traditional subject/object split between spectator and performer was challenged. Huilmand's performance and the filming of *Muurwerk* have suggested possibilities for a newly empowered feminine subjectivity evident in the notion of a contemporary *flâneuse*. The excessive performances of the three women in *Step in Time Girls*, which have challenged the culture of separate spheres that has traditionally confined women to the home, have pointed to the liberating potential of a particular kind of embodied subjectivity. In the interfaces between bodies and cities explored, possibilities for rethinking subjectivity have become apparent.

In the next chapter further mutual constructions of dancing bodies and spaces are examined in a live site specific performance set in coastal locations. The ways in which the choreography plays with particular associations that the coastal environment suggests are seen to be subversive in terms of their implications for reconstructing subjectivity.

4
Coastal Constructions in Lea Anderson's *Out on the Windy Beach*

Introduction

Beaches and the sea, like cities (see Chapter 3), are constructed spaces resonant with connotations. Seaside spaces, and the bodies that inhabit and play in them, mutually construct each other. However, this is no straightforward, simple matter. As leisure spaces, beaches and seaside resorts have associations with activities such as swimming, sunbathing and seaside entertainment. The sea is also a source of imagery in folklore and mythology and a powerful, sometimes dark, environmental force. The British postmodern choreographer, Lea Anderson, has drawn on these very different sources as inspiration for *Out on the Windy Beach* (1998) which is performed on and around a beach hut, set in a range of outdoor locations near water. The piece demonstrates the complexities and fluidities inherent in the mutual construction of subjects and spaces.

Citing the beach as a key leisure space, the French post-Marxist Henri Lefebvre suggests: 'typically, the identification of sex and sexuality, of pleasure and physical gratification, with "leisure" occurs in places specially designated for the purpose – [such as] sun-drenched beaches' (1991: 310). As a result Lefebvre claims 'such leisure spaces become eroticized' (ibid). They function as spaces of consumption in Lefebvre's terms – 'the consumption of space, sun and sea, and of . . . eroticism' (ibid: 58) – exemplified by the growth of the resort around such spaces. Resorts, products of the leisure and tourist industry, are constructed spaces with histories. The ways in which resorts and bodies construct and eroticize each other in *Out on the Windy Beach*, together with implications for the construction of subjectivity, specifically in terms of gender, are explored in the first section of this chapter.

Beaches also exist between land and sea. As shorelines, they form borders and boundaries. They are particular liminoid or in-between spaces. These coastal borderlines are explored and exploited in Anderson's choreography. In the second section of the chapter entitled 'Bodies and boundaries', the movement of dancing bodies across and between the borderlines, and the effect of the coastal environment on the bodies, are explored. These provide specific examples of the fluid nature of the mutual construction of bodies and spaces, that focuses on the limits or edges of both. In this sense there are implications for constructions of subjectivity.

Coastal environments are continually changing because of environmental factors such as erosion and pollution. This is part of their construction that appears 'natural'. These environmental changes are increasingly seen as threatening the stability of existence in a fixed, unchanging world. In some senses these changes parallel human constructions and reconstructions, evident in cloning, cosmetic surgery or body ornamentation and decoration, which can also be seen to threaten the stability and assurance of notions of fixed subjectivity. Ideas surrounding these environmental and human constructions and reconstructions are generated and fuelled by futuristic predictions on the one hand and science-fiction fantasy on the other.

Anderson explores constructions of this kind ironically in *Out on the Windy Beach*. The ways in which the American feminist Donna Haraway sees nature as constructed were referred to in Chapter 2. She claims, 'the certainty of what counts as nature – a source of insight and a promise of innocence – is undermined, probably fatally' (1990: 194). This is evident on one level in the erosion and irreversible change of the nature of coastal environments, and on another level in changes to human 'nature', both of which are referenced in *Out on the Windy Beach*. Changes in the nature of the human organism are at the centre of Haraway's manifesto for 'cyborgs', which are fabricated, futuristic, hybrids of organisms and machines, central to her 'effort to build an ironic political myth faithful to feminism, socialism and materialism.' (ibid: 190). Partly because of the parallels in Anderson's and Haraway's postmodern, ironic strategies, but also because they both draw on science-fiction generated fantasies and futurology, and because a key feature of the cyborg is 'an intimate experience of boundaries, their construction and deconstruction' (ibid: 223), there is a sense in which the dancing bodies in *Out on the Windy Beach* can be seen to be cyborgian. In this sense because Haraway sees the cyborg as 'an imaginative resource suggesting some very fruitful couplings' (ibid: 191), there are implications for ways of rethinking constructions of subjectivity. These theories, which are explored alongside *Out on the Windy Beach* throughout but specifically in the third section of the chapter entitled 'Seaside surrealism', aid an understanding of some of the more subversive elements of the dance.

Out on the Windy Beach (1998) was made for six dancers from Anderson's two companies; three from the all-female Cholmondeleys and three from the all-male Featherstonehaughs. Anderson decided to make an outdoor piece because she wanted to tap the potential for picking up new audiences who would not normally go and see her work. She likes taking dance to where people are. She began her career taking work to night club venues. In 1996 when she made *Car* set in and around a Saab car, performed in a range of outdoor public spaces, she was fascinated by the reactions of passers-by. She also said that she knew beach huts and beaches were 'a very fruitful kind of area potentially' (Anderson, 1997). She had used a beach location for part of her dance film *Cross-Channel* (1992) (see Chapter 2), and she conducted an eight day residency with dance students at a beach location in 1997 which provided her with the opportunity to begin to research and develop ideas and material for *Out on the Windy Beach*.

Out on the Windy Beach is an hour long piece accompanied by Steve Blake and Dean Brodrick playing a range of instruments, including saxaphone, bassoon, banjo, mouth harp and percussion, live on stage. They composed the accompaniment, which is inspired by a mixture of sea shanties, Amazonian Tea Dance and Appalachian Old Time music. Some of the dancers occasionally join the musicians singing or playing for their fellow performers. The set for the work, designed by Andrew White, consists of a wooden beach hut mounted centrally on the higher level of a two tiered wooden platform (Plate 3). The hut is placed at the centre back of the platform which is edged by four steps leading down to a lower level jetty-like stage, which forms a promontory about twice the width and depth of the hut. The beach hut has a veranda at the front and a central hatch-like opening in the wall, in front and below which is set a bench. The musicians play to one side of the hut on the higher platform. The set, which is painted pale sea green, is framed by a string of fairy lights and two distinctive large cerise flags, which act as markers signalling the venue from a distance. The piece, which was commissioned by the Brighton Festival, was performed at a range of waterside locations mainly around Britain, between May and August 1998.

The dancers' distinctive costumes, designed by Sandy Powell, consist of luminous, lime green, hooded, figure-hugging body suits, with straight long skirts ending in points like fish tails, over leggings that cover the feet, over which flip flops are worn. These phosphorescent outfits are complemented by reflective, orange, ski goggles and the performers' white sun-blocked lips. The effect is of weird surreal, web-footed, sci-fi aliens. When the dancers drape themselves over the edges of steps and platform, and over each other, they resemble patches of algae washed up from the sea. They become part of the landscape. They simultaneously seem to be polluting and protecting themselves from a harsh, globally warmed, sun-soaked environment. There are moments in the choreography when they look skywards as if towards some kind of otherworldly force.

The choreography consists of a series of dances for the whole company and for each of the trios of men and women. These include a mixture of unison walking towards and away from the edge of the jetty platform, dive-like poses on the edge of the jetty and various entertaining seaside dance numbers. There are also sections where the dancers lie on the platform tended and manipulated by partners and roll over, such that parts of them hang over the edge of the platform. They also enter and exit, amphibian-like, sometimes head first, sometimes feet first, through the hatch at the front of the beach hut. A host of seaside inspired images from lobster quadrilles, crab walks and mermaid poses to beauty contests are conjured up. These coexist with images of a marine invasion by surreal, sci-fi, hybrid reptilian aliens.

Towards the end of *Out on the Windy Beach* the three men appear with the bottom half of their body suits exchanged for short cerise skirts and matching knee-high boots, and the women emerge in identical cerise, short skirts, but with tattoo covered transparent tights underneath. Prior to this the dancers' costumes largely disguise their gender. This is typical of Anderson. She either uses costume and choreography to obscure gender differences (for example in parts of *Perfect Moment*, 1992 and *Smithereens*, 1999), or she emphasizes the constructed nature of gender through exaggerated differences of costume and choreography (for example in *Cross-Channel*, 1992 see Chapter 2 and Briginshaw, 1996). She is adept at foregrounding both the fluid and the constructed nature of gender in her works. In *Out on the Windy Beach* this facet of her style contributes to the revelation of the constructed and fluid nature of subjectivity that is suggested.

Eroticized bodies and spaces

A facet of the beach location that is foregrounded in *Out on the Windy Beach* is the exposure and display of the body in an erotic manner. Beaches tend to be places where people strip off to enjoy sun, sea and sand. Bodies are often displayed passively sunbathing, actively taking exercize or in some form of seaside entertainment. The dancers' bodies in *Out on the Windy Beach* are partly eroticized by their figure-hugging costumes, but these also ironically completely cover the body protecting it from the glare of the sun together with the goggles and lipblock, and militating against the display of flesh normally seen on beaches. The dancers' bodies are however eroticized by the choreography. There are several references to the 'body beautiful' in poses adopted by the performers that suggest, for example, fifties beauty queens' postures. When the three male performers emerge in their bright pink short skirts with matching boots revealing their knees, they were described by one reviewer, no doubt because they parade in a line in front of the audience, as having 'beauty queen legs glistening' (Sacks, 1998). These images are ironic references to the construction, eroticization and commodification of female beauty associated with resort beauty contests. Another ironic reference is evident when the three pink skirted males perform a version of Wilson, Kepple and Betty's thirties 'Sand Dance' which is in itself a parody of Egyptian hieratic poses. Anderson's version is a postmodern parody of a parody that has hints of faraway exotic and erotic connotations associated with the sand of the desert, rather than that of the English seaside beach.

There is a sense in which, because seaside spaces are associated with leisure and fun, their eroticization and that of the bodies that inhabit these spaces is permitted and taken for granted, it is part of their construction which is hidden. Lefebvre terms these kinds of spaces 'abstract'. He argues:

> abstract space *contains* much, but at the same time it masks (or denies) what it contains rather than indicating it. It contains specific imaginary elements: fantasy images, symbols which appear to arise from 'something else'. It contains representations derived from the established order. . . . Such 'representations' find their authority and prescriptive power in and through the space that underpins them and makes them effective. [his emphasis]
>
> (1991: 311)

Examples of representations and fantasy images which 'find their authority and power' in seaside spaces are evident in references to a range of nautical and seaside imagery in *Out on the Windy Beach*.

Poses of bathing belles seen in sailors' tattoos are referenced in the choreography and on the costumes and contribute to the eroticization of bodies, the spaces of bodies and the seaside spaces. Anderson and the dancers researched old sailor tattoos of young women in 'glamour' poses. These poses were put together to make two dances for the three Cholmondeleys in tattoed tights. References to the bathing belle postures in the dances eroticize and feminize the dancers' bodies. The same trademark sailors' tattoos of scantily clad women used as sources for the choreography were copied onto the dancers' tights (Anderson, 2000). These doubly underline the way bodies are constructed as erotic, since particular bodily spaces – the skin of the legs of the female dancers – are inscribed with the same tattoo imagery that has inspired their poses. These same images of femininity also refer to the tattoos of bathing belles normally seen on sailors' flesh, which reference the scantily clad women seen on beaches and around seaside resorts eroticizing the coastal spaces in which the dancers are performing. This performance of tattoo postures and dances underlines the ways in which bodies and seaside spaces mutually construct each other as erotic, pointing to the powerful role that gender often plays in constructing subjectivity.

When one dancer leans towards the beach hut veranda upright at an angle embracing it, the image is reminiscent of a figurehead at the prow of a ship. Postures such as this combined with the fishtail costumes in *Out on a Windy Beach* suggest mermaids. These imaginary half-human hybrid sea creatures, consisting of the head and trunk of a woman and the tail of a fish, inhabit the fantasies of myths and folklore. They are sometimes associated with the darker side of the sea as monstrous, devouring creatures who eat sailors. In this sense they share the threatening characteristics of sci-fi, aliens and cyborgs. Haraway characterizes a 'cyborg subject position' as 'dangerous and replete with the promises of monsters' (1992: 333).

Mermaids are usually seen posing on rocks, merging with the land or seascape. In this sense mermaid imagery blurs the space of the body with the space of the land or sea, the body eroticizes the landscape by becoming part of it. In Chapter 2 I discuss the ways in which landscape is traditionally treated as 'inanimate' (Oulton quoted in Lee, 1987: 23) particularly by male artists who in their search for 'beauty' and perfection see their role as transforming and subjugating 'nature', where 'nature' is traditionally constructed as feminine and 'other'. I also identify parallels between these ideas and theories of the geographer's gaze, which I suggest has parallels with the feminist film theorist Laura Mulvey's masculine gaze that objectifies 'woman'. Both objectify from a masculine perspective; the geographer's gaze objectifies the land and the masculine gaze objectifies woman. In effect the land is feminized as 'other'. The feminist geographer Gillian Rose argues, 'the female figure represents landscape, and landscape a female torso, visually in part through their pose: paintings of Woman and Nature often share the same topography of passivity and stillness' (1993b: 96). She continues: 'the sensual topography of land and skin is mapped by a gaze which is eroticized as masculine and heterosexual' (ibid: 97). This sensual topography of land and skin is evident in mermaid imagery which in its merging of the feminine and the land/seascape eroticizes both.

Mermaids, clearly gendered feminine, are part of a vast catalogue of images that contribute to the construction of femininity as, in part, erotic. In Chapter 2 these theories of the merging of 'woman' and 'nature' are raised in relation to the solo dance of a woman in water in *La Deroute* which I argue arrests the narrative in just the way Mulvey (1975) describes for female stars in mainstream Hollywood narrative films. Both the woman and the land are rendered 'transparent' and objectified by the gaze. The mermaid poses in *Out on the Windy Beach* are not transparent in this way. The costume and set render the mermaids 'out of place', they are clearly not semi-naked with flowing locks, posing on rocks. These subtle references to mermaids in *Out on the Windy Beach* demonstrate that bodies and spaces are not transparent and unproblematic. They are eroticized and resonant with cultural meanings embedded in their mutual construction. The references to mermaids, bathing belles and beauty queens, given the context of the dancers' weird, sci-fi costumes, and the performance of these traditionally gendered figures by ungendered, androgynous creatures, deliberately interrogate and challenge the ways in which the male gaze can eroticize and transparently construct bodies as landscape.

Bodies and boundaries

The beach hut set for *Out on the Windy Beach* was placed in its various performance locations within view of and facing the water. The edges of the beach hut veranda, the steps down to the platform and the edge of the platform itself all form parallel lines with the shore. The dancers foreground these land/sea boundaries in the choreography. They look towards key borders such as the shoreline and the horizon and move across, along and in between boundaries, sometimes playing with edges. Repeated back and forth movements that reflect the ebb and flow of the tides emphasize the fluidity of the boundaries.

Looking out to sea is a repeated motif emphasized by the reflective goggles of the performers. Poses adopted, such as standing on the edge of the jetty staring out to sea, and looking out whilst clasping the beach hut veranda upright pole, foreground the look. The goggles make the looks appear ominous. The traditional spectator/performer relationship where the former objectifies the latter is reversed. The performers return spectators' gazes reflected in their eye wear. By doing this they cross the traditional boundary between performer and spectators, disrupt the conventional construction process which involves objectification and render the construction instead a two-way process, which is mutual and reciprocal. They are problematizing the subject/object binary bound up with traditional constructions of subjectivity and foregrounding alternative mutual construction processes. The problematization of binary oppositions such as that between subject and object is a key characteristic of Haraway's cyborgs. She says of cyborg imagery that it 'can suggest a way out of the maze of dualisms in which we have explained our bodies and our tools to ourselves' (1990: 223) and that, 'cyborg figures... are the offspring of implosions of subjects and objects and of the natural and artificial' (1997: 12). The dancers in *Out on the Windy Beach*, because of their goggled looks towards the sea and spectators, blur the subject/object binary. They can be seen to be 'cyborgian offsprings of implosions of subjects and objects'. They also blur the natural/artificial binary with their references to mermaids, because of the radical juxtaposition of their mermaid poses with their surreal, sci-fi costumes.

Land/sea boundaries are foregrounded in the choreography by performers moving across, around and in between them and playing on the edges. The artificiality of these spatial limits is highlighted when performers teeter on or hang over the edges of them. There are several examples of dramatic encounters with the edge of the jetty-like platform in *Out on the Windy Beach*. Dancers repeatedly walk to the very edge and make large gestures upwards and outwards with outstretched arms and open chests before turning to retreat back across the platform towards the hut and land. They lie on the platform with their heads over the edge when on their backs or sides, having rolled to the periphery.

They never leave the platform by moving onto the ground in front of it, two or three feet below them, instead by their gestures and performance of diving poses, they ironically give the impression of water below. The connotations of the limits and constraints of this particular boundary or borderline, as well as its ultimate artificiality and constructed nature, are highlighted by the choreography and performance. At times it is as if the dancers are riskily flirting with this boundary in a cyborgian manner. Haraway claims, 'my cyborg myth is about transgressed boundaries, potent fusions and dangerous possibilities' (1990: 196).

The dancers' dramatic exits and entries through the beach hut hatch, head or feet first, slithering through the opening, foreground this boundary and, in this instance, its unruly transgression. The manner of entrance and exit is so unusual that it attracts attention and further highlights the constructed nature of the boundary being crossed and the bodies performing. These actions are decidedly not natural, the boundary transgression is so bizarre it suggests possible 'potent fusions and dangerous possibilities' in the manner of Haraway's cyborgs.

Boundaries or borders are also reiterated by lines of dancers in the choreography. Dancers sit in a line on the veranda bench, and on the steps that lead down to the jetty from the hut, they stand in a line at the edge of the jetty, and pose in racing dive positions on the same jetty edge. The distinctive land/sea border is thus repeatedly stated. The continual crossings and recrossings of these stated boundaries, while emphasizing their constructed nature, also point to their fluidity and that of the boundaries of the performers. If boundaries of space and bodies are seen as constructed and fluid in this way, then they can be open to possibilities for change and reconstruction.

This fluidity is particularly evident in patterns of back and forth travel across the jetty-like platform. If the performers are not advancing or retreating towards or away from the water, they are, for much of the time, gesturing towards and away from it. The ebb and flow of the choreography titillates and teases, playing with the 'no-man's land' of the jetty perched between sea and shore. This spatial dimension of the choreography, predominantly towards and away from the beach hut and the sea or water, foregrounds the construction of this space as special. It marks the place where the land ends and the sea begins. The continual movement back and forth and looking out to sea which dominate in the choreography, as well as the nature of the movements performed, emphasize the socially constructed nature of the space. The choreography also foregrounds and plays with the actual fluidity of the borderlines between land and sea which continually change with the tide. Horizontal lines of performers walk in unison back and forth across the platform towards and away from the water repeatedly, sometimes accompanied by raised outstretched arms broken at the wrists with hands pointing down. These simple gestures are almost cartoon-like, they look like pastiches of children's drawings of birds. The simplicity of the back and forth walking and the bird-like arm gestures which occur towards the beginning of the dance set the scene for what is to follow. Simple movement statements and gestures indicate the influence or effect of the coastal environment on the choreography and the performers. At the same time these are not subtle connections with tides and sea birds as for example in Merce Cunningham's dance, *Beach Birds* (1992). The movements and gestures are crude and obvious, performed in a self-reflexively postmodern, ironic, manner typical of Anderson's choreography (see Briginshaw, 1996). Their performance deliberately exposes their artificial or constructed nature while also revealing ways in which seaside spaces are constructed by cultural associations.

These excerpts emphasize the range of borderlines and boundaries of bodies and space that can be explored in this coastal environment. The choreography foregrounds the constructed nature of spatial borderlines where land meets sea. It reveals them to be cultural constructions; lines drawn on maps to mark out territory, but it also shows, often in a cyborgian manner, how the fixity can be broken and boundaries blurred, allowing potential for change through fluidity. Another example of the fluidity of boundaries, but this time bodily boundaries, is evident in the references to mermaids, who are creatures whose bodily boundaries are necessarily blurred. The implications for subjectivity of blurred bodily and spatial boundaries, which can suggest blurred conceptual boundaries, are taken up in the next section.

Seaside surrealism

The lurid, phosphorescent, lime green, figure-hugging body suits that the dancers wear in *Out on the Windy Beach* give the performers an unnatural and surreal appearance. The dancers stand out futuristically against the faded pale sea green paint of the beach hut set. Their costumes are almost like second skins, they eroticize the dancers' bodies but they also make them look horrifically alien, like sci-fi extra-terrestrials. The goggles and lip-block together with the full body coverage and luminosity of the costumes, also suggest hyper-protective gear worn by environmental activists. These polysemic creatures that embody a range of possibilities are unstable and difficult to fix, they perplex and provoke in a hybrid, cyborgian manner. The hybridity of cyborgian creatures is central to their constitution as blurrers of boundaries and as blurred and fluctuating subjects and identities. The ways in which cyborgs and the dancers blur binaries such as subject/object and natural/artificial have already been identified. Haraway suggests, 'cyborg anthropology attempts to refigure provocatively the border relations among specific humans, other organisms and machines' (1997: 52). When the surreally costumed dancers in *Out on the Windy Beach* crawl on all fours, walk crab-like across the space on hands and feet, lie with their heads hanging over the edge of the platform or all stare menacingly out to sea, they certainly appear to be provocatively refiguring 'the border relations among specific humans and other organisms' in a cyborgian fashion.

The dancers are often seen on the ground lying, rolling or curled up, and because they perform at the water's edge, they sometimes resemble marine deposits washed up with the tide. When they lie with their heads hanging over the edge of the jetty they transgress the border messily, soiling it like pollutants. Anderson has said of the piece 'there was a lot of "futurology" stuff – reading reports about what will happen to land and sea in fifty years time – evolution and extinction' (2000). Ecological factors play an important part in Haraway's theories of cyborgs. She has said 'cyborg life forms that inhabit the recently congealed planet Earth . . . gestated in a historically specific technoscientific womb' (1997: 12). One example of an arena for such life forms is, 'that planetary habitat space called the ecosystem, with its . . . birth pangs in resource management practices in such institutions as national fisheries . . . and with its diplomatic forms played out in 1992 at the Earth Summit in Rio de Janiero' (ibid: 13). References to coastal environmental pollutants in the costumes and choreography of *Out on the Windy Beach* ally the alien forms of the dancers' bodies even more with cyborgs.

A key border that Haraway and Anderson both flout in their work is that of gender. Haraway proposes that her manifesto is an 'effort to contribute to socialist – feminist culture and theory . . . in the utopian tradition of imagining a world without gender' (1990: 192). Sandy Powell's all-in-one, hooded, unisex bodysuits cover everything except the dancers' faces which are goggled, and render them androgynous. The performers' androgyny is further underlined by the choreography, since for much of the time it does not discriminate according to gender. There is much dancing in unison often in lines with a strange focus on tiny gestures typical of Anderson's choreographic style. There is an unstoppable mathematical flow in the dance and music resulting in a continuous stream of dancing, which blurs genders.

Gender blurring also occurs when the dancers' bodies sinuously intertwine in bizarre, mutating, mating rituals. It is difficult to tell one body from another as they bend, twist, slither and slide around and over each other, mutating into different shapes and forms. The performance is erotic and alluring, yet because of the costumes and goggles, it is also distanced and disturbing. Anderson said that environmental research into the future of land/sea borders was the starting point for the mating games in the choreography (2000). These mutations have resonances with Haraway's cyborgs whose temporal modality is 'condensation, fusion, and implosion' (1997: 12). She has also claimed 'cyborgs signal disturbingly and pleasurably tight coupling. Bestiality has a new status in this cycle of marriage exchange' (1990: 193). This description could apply to the coupling creatures in *Out on the Windy Beach*. Mutating and multiplying in this way can have associations with the subversive characteristics of pollution, which Haraway reclaims for positive political use. She argues 'cyborg politics insist on noise and advocate pollution, rejoicing in the illegitimate fusions of animal and machine' (ibid: 218).

In this sense, the fusions evident in the mutating couples in *Out on the Windy Beach* can be interpreted in terms of positive regeneration. Haraway suggests, 'cyborgs have more to do with regeneration and are suspicious of the reproductive matrix and of most birthing' (1990: 223). In other words, conceptually she privileges a mutating form of renewal and regeneration over traditional birth. Mutation as a means of explaining change has connotations of multiplicity over the more linear notion of evolution which has associations with progressive and purposive change, which it has been suggested is 'found throughout Enlightenment thought' (Gisbourne, 1998).[1] Haraway states, 'I would . . . like to displace the terminology of reproduction with that of generation. Very rarely does anything really get *reproduced*: what's going on is much more polymorphous than that' [original emphasis] (1992: 299). The mating couples in *Out on the Windy Beach*, as they roll over each other and entwine themselves in fluid, amoebic-like connections where many different body parts come into contact, are certainly polymorphous, often resembling creatures such as salamanders, which Haraway cites. She indicates that for them 'regeneration after injury, such as the loss of a limb, involves regrowth of structure and restoration of function with the constant possibility of twinning or other odd topographical reproductions at the site of former injury. The regrown limb can be monstrous, duplicated, potent.' (ibid: 223). Suggestions of the monstrous, androgynous, regenerated, new beings Haraway imagines and theorizes are evident in the choreography in *Out on the Windy Beach*.[2]

References to 'regeneration' and a 'monstrous world without gender' also have resonances with the Russian literary critic Mikhail Bakhtin's ideas of the carnivalesque (see Chapter 9). There are carnivalesque elements in *Out on the Windy Beach*, particularly in the second half when the dancers further transform themselves through costume changes, exchanging their leggings and fishtail long skirts for short, above-the-knee, cerise skirts. In this part of the performance various seaside and popular variety entertainments are referenced; beauty contests and numbers such as the thirties Wilson, Kepple and Betty 'Sand Dance', which the men perform accompanied by the women sitting on the beach hut bench playing miniature guitars. This is an example of a gender reversal typical of Anderson, having the men dance while the women play in the band. Anderson has said of her all-female company's early performances in night clubs that she wanted to subvert and get away from the use of women as dancing display objects accompanying the all-male music bands (Adair, 1987). One strategy was to make a piece, *Baby, Baby, Baby* (1986), which parodied and satirized female backing groups like the Supremes.

The inclusion of the Wilson, Kepple and Betty 'Sand Dance' and the beauty contest references in *Out on the Windy Beach* is evidence of Anderson's 'everlasting fascination with Music Hall'and popular variety acts and of her desire to include an 'end of the pier' aspect (Anderson, 2000). These elements are also further evidence of the ways in which seaside spaces such as resorts are eroticized and commodified and are, in Lefebvre's words, 'spaces of consumption' (1991: 58). Commodification and consumption are defining features of postmodernity. Haraway recognizes this, stating that 'perhaps cyborgs inhabit less the domains of "life," with its developmental and organic temporalities than of "life itself" ... [which] is enterprised up, where ... the species becomes the brand name and the figure becomes the price' (1997: 12). Resorts and entertainment have histories as spaces and modes of consumption. The beauty contests and the Wilson, Kepple and Betty 'Sand Dance' reference the histories of resorts and popular entertainment respectively. However, given the context and costuming, these dances ironically and surreally hint at parallels between seaside and music hall entertainment and the more futuristic kind of similarly, popular entertainment evident in contemporary sci-fi movies. Haraway suggests that 'cyborg spatialization' is about '"the global"' and that 'the globalization of the world ... is a semiotic-material production of some forms of life rather than others' (ibid). Examples of 'forms of life' that 'craft the world as ... global', she argues, are 'cyborgian entertainment events such as *Star Trek, Blade Runner, Terminator, Alien* and their proliferating sequelae ... embedded in transnational, U.S.-dominated ... media conglomerates, such as those forged by the mergers of the Disney universe with Capital Cities' (ibid: 13). The performance of seaside entertainment numbers and the Sand Dance by sci-fi, extra-terrestrial, alien figures in *Out on the Windy Beach* is both surreal and carnivalesque in its blurring of boundaries between human and non-human and its conjunction of entertainment references from different historical periods. This bizarre juxtaposition is mischievous having suggestions of the trickster figures of carnival which Haraway also associates with the cyborg. She proposes, 'in the 1990s, across the former divide between subjects and objects and between the living and nonliving, meaning-in-the-making ... is a more cyborg, coyote, trickster, local, open-ended, heterogeneous, and provisional affair' (ibid: 127).

The mischievous inclusion of these carnivalesque entertainment numbers in *Out on the Windy Beach* given the costumes – the dancers are still wearing their goggles and lipblock – is heavily ironic. As Haraway indicates, 'a cyborg body ... takes irony for granted' (1990: 222) and 'the cyborg is resolutely committed to partiality, irony, intimacy, and perversity. It is oppositional, utopian, and completely without innocence' (ibid: 192). There is certainly an irony in Anderson's transformation of potentially alienating, monstrous, sci-fi creatures into sexy, alluring entertainers who are fun.

The utopian characteristics of Haraway's theories lead her to posit new possibilities for subjectivity. She draws on the imaginary worlds of science fiction in a positive way. She writes of 'moving ... to a science fictional, speculative factual, SF place called, simply, elsewhere' (ibid: 295). It is in this sense that she is making suggestions for new ways of thinking beyond traditional notions of the subject. She argues, 'the subject is being changed relentlessly in the late twentieth century' (ibid). These new ways of thinking beyond traditional notions of the subject involve connection and embodiment. She is concerned 'to produce ... effects of connection, of embodiment, and of responsibility for an imagined elsewhere' (ibid). The embodied interconnectivity of the mutating, mating dances in *Out on the Windy Beach* has already been described. At another point when the dancers are in pairs, one of them lies on the ground, whilst their partner leans over them, lifting and lowering limbs and placing hands on them. They look as if they are tending and caring for them, perhaps suggesting the kind of supportive and responsible connections Haraway posits.

Anderson's blurring of a range of boundaries in her choreography has already been mentioned in connection with Haraway's references to blurred binaries. Haraway argues, 'science fiction is generally concerned with the interpenetration of boundaries between problematic selves and unexpected others and with the exploration of possible worlds' (1992: 300). By bringing sci-fi elements into a postmodern dance piece Anderson has brought together science and culture, blurring a key binary. Haraway also sees her own work as rooted in the premise that 'science is culture' (ibid: 296). Central to Haraway's cyborgian vision for the future is the appropriation of science and new technologies, particularly communication technologies, for her own feminist ends. She sees these technologies playing an important part in rethought, embodied subjectivities. She claims 'communications technologies and biotechnologies are the crucial tools recrafting our bodies' (1990: 205), and 'myth and tool mutually constitute each other' (ibid: 206). This mutual construction parallels that of bodies and seaside spaces evident in *Out on the Windy Beach*. These elements have been seen to eroticize each other, but they also, by their juxtaposition, have been seen in a subversive and ironic cyborgian manner, to construct each other as surreal.

Conclusion

There are several ways in which the choreography of *Out on the Windy Beach* explores and plays with interfaces between bodies and coastal environments. The bodies and spaces in *Out on the Windy Beach* have been seen to eroticize each other through the inclusion of images of mermaids, bathing belles and beauty queens, and of popular, entertaining dance numbers. Interfaces between bodies and coastal spaces are also foregrounded and played with in the choreography, which emphasizes the land/sea spatial dimension, through the continual reiterating and crossing of borderlines and boundaries. Spaces and bodies have been seen to mutually construct each other, with gender at times informing the construction process. Importantly the dances also foreground the fluid nature of the constructed boundaries of bodies and spaces, indicating that these constructions are not fixed but open to change, allowing for new possibilities concerning the construction of subjectivity.

These new possibilities become more apparent with reference to Haraway's theories of cyborgs, which when explored alongside *Out on the Windy Beach* reveal some of its more subversive elements. For example, the polluting powers of Anderson's mutating, alien, coastal creatures, because of the ways in which they are seen to blur fixed conceptual binaries and boundaries, can be seen as positive and regenerative when read in Haraway's terms. Anderson's witty and ironic animation of her 'patches of algae' in end of the pier dance routines is a graphic illustration of such regeneration. The political impact of *Out on the Windy Beach* in terms of its interrogations of the fixity of gender is more apparent when the various antics of the performers are seen as cyborgian. As Haraway argues, 'cyborg monsters in feminist science fiction define quite different political possibilities and limits from those proposed by the mundane fiction of Man and Woman' (1990: 222). Initially the import of the links with ecology and futurology in *Out on the Windy Beach* is difficult to fathom, but when the links are explored alongside Haraway's theories, the ways in which they suggest new possibilities for rethinking subjectivity in an embodied, unfixed, polymorphous and more connected way are revealed.

Part II
Dancing in the 'In-Between Spaces'

5
Desire Spatialized Differently in Dances that can be Read as Lesbian

Introduction

While the discussions of dance works in Part I were concerned with constructions, dynamics and qualities of particular outdoor spaces and locations, and how these contribute to constructions of subjectivity, specifically with reference to gender, the next three chapters focus on virtual or metaphorical spaces seen as in-between. These in-between spaces existing on the borderlines of body/space interfaces and spatial interfaces are explored specifically with reference to sexuality in this chapter, and cultural differences in the next two chapters, to see what potential they might hold for rethinking and challenging traditional constructions of subjectivity.

This chapter explores the spatialization of desire in postmodern dances that can be read as lesbian. It focuses on the interfaces between bodies, and bodies and space, in particular kinds of in-between spaces. I have chosen to look at these dances because of the potential a lesbian reading can suggest for rethinking desire and conceiving it as spatialized differently. It is important to clarify that I am not using the term 'lesbian' to describe an identity rooted and fixed in the body. My position, alongside that of the feminist Cathy Griggers, is that bodies, including lesbian bodies, 'only exist in a process of constant historical transformation...there are only hybrid bodies, moving bodies, migrant bodies, becoming bodies' (1994: 128).

Theories of desire refigured are explored in the work of the Australian feminist Elizabeth Grosz, the French poststructuralists Gilles Deleuze and Felix Guattari and the French feminist Luce Irigaray, who all argue for a productive notion of desire based on touching, sensation and interfaces, rather than on lack. Grosz, in her essay entitled 'Refiguring

Lesbian Desire' explicitly states her aim to 'think desire beyond the logic of lack' (1994a: 69). Both the Cartesian notion of a material body and immaterial mind or spirit, and the Freudian notion of a normatively heterosexual desire predicated on a perceived lack, which requires an other, who is marked as different, are based on binary oppositions. Desire based on lack is characterized by binary oppositions of not only self/other, but also subject/object, active/passive and presence/absence. This chapter argues that non-dualistic ways of thinking which blur binary oppositions offer ways of theorizing theatre dance to reveal the potential in it for refigured desire. Grosz describes desire as lack in spatial terms as 'an absence, lack, or hole, an abyss seeking to be engulfed, stuffed to satisfaction' (ibid: 71). In these descriptions space is conceived as empty, unproductive, a void waiting to be filled. In the theories of Grosz, Deleuze and Guattari, and Irigaray, and in the dances examined here, I argue that space, desire and bodies need to be rethought as reciprocally productive, that is that they continually produce and are produced by each other.

After briefly outlining key concepts such as assemblages and becoming (from Deleuze and Guattari), which inform Grosz's notions of refigured desire, and the imagery of two lips (from Irigaray), the spatial configurations associated with these are employed to describe interfaces between bodies, and bodies and space in two dance video duets and two solos. These spatial configurations all focus on the productive interconnectivity between bodies and space, allowing space to be seen as discursively constructed and sexualized, and subjectivity to be rethought in terms of a fluid notion of becoming rather than a fixed idea of being. *Reservaat* and *Between/Outside* were each made as dance films for two female performers, respectively in the Netherlands in 1988 and in London in 1999. *Virginia Minx at Play* and *Homeward Bound* are both solos danced by the choreographers and first performed at the Riley Theatre, Leeds in 1993 and at Chisenhale Dance Space, London in 1997 respectively.

Lesbian desire refigured

In her essay 'Refiguring Lesbian Desire' Grosz suggests that desire when seen 'primarily as production rather than ... lack' (1994a: 74) can be conceived in terms of 'the energy that creates things, makes alliances and forges interactions between things' (ibid: 75). She draws on the work of Deleuze and Guattari because they see desire as 'immanent, positive, and productive, as inherently full' (ibid) and 'as ... what connects, what makes machinic alliances' (Grosz, 1994b: 165). An alliance in their terms is a pact 'an infection or epidemic' (Deleuze and Guattari, 1988: 247). In this sense it is dynamic and everchanging, it has the ability to spread. 'The ways in which . . . bodies come together with or align themselves to other things produce what Deleuze has called a machine: a nontotalized collection or assemblage of heterogeneous elements and materials' (Grosz, 1994b: 120). These machinic connections are characterized by 'intensities' and 'multiplicities'. Importantly an assemblage does not describe what things are in terms of fixed states of being, but what things do, how they behave, and how they are structured and relate in terms of processes, it is dynamic. As Deleuze and Guattari state, 'it is a composition of speeds and affects . . . a symbiosis' (1988: 258). ' In this sense Deleuze and Guattari's work 'provide[s] an altogether different way of understanding the body in its connections with other bodies . . . it is understood . . . in terms of . . . the linkages it establishes, the transformations and becomings it undergoes' (Grosz, 1994b: 164–5). The value of their notion of 'becomings' is that they 'involve destabilizing recognizable patterns of organization' and they 'indicate[s] new possibilities in self-transformation' (Lorraine, 1999: 181). They can be particularly pertinent when considering the different spatialization of desire as expressed in dance because they involve 'challenging conventional body boundaries' (ibid: 183).

The different spatialization of lesbian desire is identified by the British feminist Penny Florence, in her discussion of sex and spatial structure in a Manet painting (*Mlle Victorine en costume d'Espada*, 1862). She suggests that the little the French psychoanalyst Jacques Lacan has to say about lesbians 'hints towards a desire which is *spatialized differently* from its articulation in most of his writing' (emphasis added) (1998: 262). She continues: 'Rather than a "contorted" space, manifest in intervals this female desire is contiguous, with no loss and no gap, but rather a touching' (ibid : 263). As Florence comments, these ideas have strong resonances with Irigaray's image of two lips, where the contiguity evident permits a plural sexuality (Irigaray, 1985b: 28), 'a multiplicity which allows for subject-to-subject relations between women' (Whitford, 1991a: 182). It is important to recognize that Irigaray is not referring in any literal sense to the female anatomy, although that is the origin of the image, which is rather used as a discursive strategy. The image works metonymically through contiguity, association or touching (ibid: 180) which allows for *process*. This emphasis on process has similarities with Deleuze and Guattari's notion of becoming. Irigaray claims: 'Woman is neither open nor closed. She is indefinite, in-finite, *form is never complete in her....* This incompleteness in her form... allows her continually to become something else' (her emphasis) (1985a: 229), 'metamorphoses occur... transmutations occur' (ibid: 233). The image is particularly relevant for exploring spatial configurations of lesbian desire in dance because, according to Irigaray, it focuses on women's limits or 'edges touching each other' (ibid: 232–3), which 'do not constitute those of a body or an envelope, but the living edges of flesh opening' (quoted in Whitford 1991a: 162). I see this as a fluid ever changing body/space interface, where 'there is something which exceeds all attempts to confine/define her [woman] within a system of discourse' (Whitford, 1991b: 27). The excess, openness and fluidity stem from the multiplicity and polymorphous perversity of woman's sexuality. Irigaray claims: '*woman has sex organs more or less everywhere. She finds pleasure almost anywhere... the geography of her pleasure is far more diversified, more multiple in its differences, more complex, more subtle, than is commonly imagined*' (her emphasis) (1985b: 28). This suggests 'a mode of being "in touch" that differs from the phallic mode of discourse' (Burke quoted in Whitford, 1991a: 172).

From these sources there is the suggestion that desire can be seen to be spatialized differently. This different spatialization is not based on lack, or space seen as distance, but rather on surfaces, intensities, interfaces and touching. Images of machinic connections and the 'two lips' with associations of becoming, transformations and multiplicities, which blur boundaries challenging binary oppositions, assist in describing and characterizing this spatialization.

The performance of gender and sexuality

When considering the choreography of refigured desire it is important to recognize that dancing bodies are constructed as gendered and sexed. The body is always already gendered and conventionally sexualized as heterosexual. As the American queer theorist Judith Butler argues, 'the association of a natural sex with a discrete gender and with an ostensibly natural "attraction" to the opposing sex/gender is an unnatural conjunction of cultural constructs' (1990: 275). Butler claims that genders of bodies are *performed*, repeatedly performed in social and cultural performative acts. However, it is important to distinguish between this kind of performance or performativity, and dance performance. The performance of gender is something that everyone does repeatedly, usually without realizing it is a performance. It is part of what it means to inhabit a gendered identity and be a meaningful subject. This is different from dance performance which is recognized as such. As Butler suggests, performativity is different from theatrical performance because performativity 'consists in a reiteration of norms which precede, constrain, and exceed the performer' (Butler, 1993: 234). Within dance performance the performativity of gender and sexuality is hidden or so apparent it is invisible. It exists alongside and interacts with the dance performance. Investigation of these interactions between dance performance and performativity in dances that can be read as lesbian, has revealed that there is an active seductive energy that exists in the spaces between dancers (Briginshaw, 1998). As the French post-Marxist Henri Lefebvre claims when discussing the spatial effects of bodily rhythms, which in part may be 'distilled into desire', 'an animated space comes into being which is the extension of the space of bodies' (1991: 205–7). It is this animated space that I explore further in the dances analyzed here. My analysis, following Butler, is based on the premise that bodies are discursively constructed as gendered and sexed through repeated performative acts.

The Dances – *Reservaat* (1988), *Between/Outside* (1999), *Virginia Minx at Play* (1993) and *Homeward Bound* (1997)

Reservaat, made by Clara Van Gool in 1988 and filmed in black and white, is set in a country park or reservation with trees and a lake, where Martine Berghuijs and Pépé Smit dance a tango (see cover photo).[1] It has been screened at various lesbian and gay film festivals in North America and Europe throughout the nineties.[2] *Between/Outside,* filmed and choreographed by Lucille Power in 1999, in which she performs with Sarah Spanton, is set in a concrete stairwell of a block of London flats. Both dance films last about seven minutes and are accompanied by the sounds of the performers: the crushing of leaves underfoot by the tango dancers in *Reservaat*, and the dancers' echoey footsteps on the uncarpetted stairs in *Between/Outside*. Whereas tango dance steps constitute the choreography in *Reservaat,* pedestrian movement, mainly walking up and down stairs and sitting, is used in *Between/Outside*. In both, the women are always close to each other and touch, in *Reservaat* maintaining the tango embrace throughout. The intensity that the proximity of the dancers creates is a key focus in both works. I explore the spatial configurations in the interfaces between the dancers and their implications for refiguring desire. The 'Watch me witch you' section (10 mins) from *Virginia Minx at Play* (77 mins) choreographed by Emilyn Claid in 1993, and *Homeward Bound* (33 mins) choreographed by Sarah Spanton in 1997 are danced on bare stages. They both create androgynous characters, a Latin American dancer and a sailor respectively (Plates 4 and 5). The choreography, costumes and music are crucial in character construction. In *Virginia Minx at Play,* Claid, performing virtuoso Latin dance vocabulary, wears black, close-fitting, high-waisted, silk trousers, a deep pink sequined fringed waistcoat, matching pink long satin gloves and a black velour fedora hat. She is accompanied by a Latin style song entitled 'Sombrero', sung live on stage by a female musician, Heather Joyce. In *Homeward Bound,* Spanton performs a mixture of traditional nautical dance steps such as the sailor's hornpipe and other movements that evoke the swell of the ocean. She is dressed in a white sailor suit with blue square collar, and she has a lilo as a single prop. She dances partly in silence but mainly to recordings of eleven sea shanties sung by male folk singers in both the male and female third and first person. In both works tension builds as the spaces between parts of the dancers' bodies and between the performers and spectators are played with in performance.

The cultural contexts of the tango in *Reservaat* and of the broader Latin dance conventions in *Virginia Minx at Play* inscribe the dancing bodies in particular ways. As Grosz, following Foucault and Butler, argues, 'bodily materiality... through corporeal inscriptions... is constituted as a distinctive body... performing... in socially specified ways' (1994b: 118). Bodies perform in distinctively gendered and sexualized ways and these performances are often underlined and embellished with other conventional performances, such as those of tango and Latin American dance. Grosz continues: 'Bodies are fictionalized... positioned by various cultural narratives and discourses, which are themselves embodiments of culturally established canons, norms, and representational forms... they can be seen as living narratives' (ibid). As the dance theorist Marta Savigliano demonstrates, tango is a rich living cultural narrative with its own canons, norms and contesting representational forms. Like many Latin dance styles it is imbued with associations of the erotic and exotic and its rendering of heterosexual desire masks the many tensions in its narrative that Savigliano reveals. This provides an important context for readings of *Reservaat* and *Virginia Minx at Play* in the sense that the narratives and discourses of Latin dance conventions construct the dancing bodies in these two works initially as contained, and part of the traditional heterosexual narrative of desire as lack. It is only when these dance forms are played with in specific ways in the performances, that possibilities for a more productive notion of desire spatialized differently become apparent.

Surfaces in contact

Reservaat opens with a shot of a lake surrounded by trees; the reservation of the title, which sets the scene for what is to follow: a close-up shot of fur, with a hand, fingers spread, gradually coming into view, moving slowly across it. As the fur begins to turn, it becomes clear that it clothes a woman's back, held in a tango embrace by another woman, dressed in an almost identical, but darker, fur costume. The camera lingers on the hand/fur interface sensually, attention is focused on the touch and textures by the close-up.

In *Between/Outside* there is a similar moment of contact that occurs towards the end of the piece when the two women who are dressed rather drably – Lucille Power in a grey leather coat and Sarah Spanton in a red imitation fur jacket – are seen in profile. Spanton, standing behind Power, lifts Power's shoulder length hair out from under her coat collar and smooths it down once with her hand. While this is happening, Power closes and opens her eyes twice, possibly blinking, or possibly briefly closing her eyes in pleasure, cherishing what is happening. Soon after, the women's feet pass each other on a single stair, as if on a narrow ledge – another moment of proximity. 'What at one moment is suggested as an impersonal relationship is the next disrupted when a different level of intimacy is revealed' (Power on *Between/Outside*, 1999).

The shot of Spanton's hand smoothing Power's hair is a touch that is in fact the culmination of a series of increasingly close encounters between the two women. It can be compared to the opening shot of the hand on fur in *Reservaat,* which sets the scene for the close bodily contact that follows. These touches, placed as they are at the beginning and end of the two works, are significant. They stand out in the choreography because of the lengthy, close-up filming in *Reservaat* and the context in *Between/Outside*; up until this point the two women's relationship has been ambiguous, and this touch seems to signify a level of intimacy not evident earlier. In both cases body parts in contact and textures are key. It is the hand that touches, and it touches something soft and silky, on the other woman's body. As Grosz suggests:

> we must focus on the elements, the parts . . . In looking at the interlocking of two such parts – fingers and velvet . . . there is not as psychoanalysis suggests, a predesignated erotogenic zone . . . Rather, the coming together of two surfaces produces a tracing that imbues both of them with eros or libido, making bits of bodies . . . parts, or particular surfaces throb, intensify, for their own sake.
>
> (1994a: 78)

Intensities are suggested in both pieces by these moments. The lengthy shot of the hand on fur is an instance of, to paraphrase Grosz, an encounter or interface between one body part and another which produces an erotogenic surface and lingers on and around it for evanescent effects (ibid: 78). These touches close the gap between the women which, as Irigaray has stated, allows for a transmutation or metamorphosis to occur, for the possibility of becoming something else. The 'something else', Irigaray suggests, is an interconnectivity of the two bodies: 'I'm touching you, that's quite enough to let me know that you are my body' (1985b: 208). This interconnectivity can be compared to the assemblages that Deleuze and Guattari theorize which are discussed later in the chapter.

In *Virginia Minx at Play* and in *Homeward Bound* there are more examples of touches of smooth surfaces of fabric or hair covering bodies. But this time the touches are self touches. Claid deliciously fingers her own pink satin gloved palm – pink silk on pink silk – she draws suggestive circles in it before leading our gaze up her arm with her finger which ends erotically in her mouth, from slippery silk over bare flesh to fluid. She frequently strokes her own crotch in an affectionate and sometimes also in a more masturbatory fashion enjoying her own body, her satin gloved hand sliding easily over the front of her black silk trousers.

The programme notes for Sarah Spanton's performance of *Homeward Bound* (1997) state, 'Sarah Spanton explores her bisexual identity by taking a personal voyage of discovery' and 'saucy wink ship's mate footpump lilo rubber inflate deflate inhale exhale flirt sailor – "boy" toying cruising ... seducing "passing" hello sailor'. She begins by inflating a lilo with a footpump looking seductively at the audience as she, at first slowly, and then with increasing speed and intensity, pumps with her foot, whilst also swaying her hips and body. The build up in speed and tension of the 'heavy breathing' of the lilo hints at orgasm. This is the first of several titillations with the audience which often involve Spanton seductively swaying her hips from side to side sometimes putting her hands on her hips, both on one, or one on each, while cheekily approaching the audience and then retreating. For much of the time Spanton stares out boldly at the audience – she flirts with her body and her eyes playing with the distance between herself and us. She circles her arms and shoulders, runs her fingers through her cropped boyish haircut suggestively and continually plays on the swell and wave-like body movements that suggest both sea and sex.

Both Spanton and Claid in their performances are in very different ways enjoying sensually touching their own bodies – smooth surfaces feature again – hair and silk. Irigaray has suggested: 'When I touch myself I am surely remembering you' (1985b: 215). Importantly for Irigaray this 'you' is another woman, partly because she sees the power of touch coming from the original sensation of connection with the mother in the womb. She argues, 'the singularity of the body and the flesh of the feminine comes... from the fact that the sensible which is the feminine touches the sensible from which he or she emerges. The woman being woman and potentially mother... the two lips... can touch themselves *in* her... these two dimensions... are *in* her body. And hence she experiences it as volume in a different way?' (her emphasis) (1993: 166).

Becoming/transformations in *Reservaat* and *Homeward Bound*

After the opening glimpses of the hand on fur in *Reservaat,* the shot cuts to the dancers' intertwined legs in knee length dark suede boots and black opaque tights, performing a *sandwich* (or *mordida* when one partner's foot – usually the woman's – is sandwiched between those of her partner – usually the man). They then turn and walk sideways, mirroring each other, in typical tango style, ending with an embellishment. Only the sounds of the wind and what could be a bird in the distance are heard. When the legs disappear, the camera lingers on a shot of the undergrowth. In these opening moments the focus is on connections between *parts* of bodies – hand on fur and intertwined legs – and on the reservation environment. An interconnectivity between the parts of the two bodies, and between them and the surrounding environment is immediately suggested. Grosz claims: 'to use the machinic connections a body part forms with another... is to see desire and sexuality as productive... a truly nomad desire unfettered by anything external for anything can form part of its circuit and be absorbed into its operations' (1994a: 79). In *Reservaat* I am suggesting that the habitat of the reservation forms part of the circuit of desire and is absorbed into its operations. The emphasis in the filming and choreography seems to be on the two dancers becoming one with each other and with the environment. Grosz describes the Deleuzian notion of becoming as, 'entering into a relation with a third term and with it to form a machine' (ibid: 78). In *Reservaat* the habitat environment of the reservation could be seen as representing a third term in a machinic transformation with the two dancers, and in *Between/ Outside,* the dull and empty, yet containing, stairwell might also be seen in this role. It provides a space for the two women's brief encounters and its ordinariness enhances the encounters' frisson, making them seem unlikely.

In *Reservaat*, a few minutes into the film the two dancers are seen, heads and shoulders only, slowly turning as they travel in their tango embrace, they look around, at and past each other. The shots jump quickly from a dancer's foot noisily crushing a twig, to a fox looking out of a hole and howling, then to a close-up of a dancer's eyes turning to look, presumably at the fox, or for the source of the sound. After close-ups of her partner's eyes and ear, and her own eyes, the dancers look back at each other and, maintaining their tango embrace, jump together. There follows a pattern of steps in the *milonga* style ending in an embellishment. The fox is then seen disappearing down a hole and the couple dance off in the distance.

The dance is clearly not presented as a 'normal' tango. The setting, costume and lack of tango music, as well as the performers' gender, suggest otherwise. It seems more likely that the machinic connection of the tango embrace, given the setting, the fur costumes and the juxtaposition with a fox in the filming, could be read as a Deleuzian instance of 'becoming animal'. Making reference to Deleuzian notions of becoming, Grosz argues:

> One 'thing' transmutes into another, becomes something else through its connections with something or someone outside ... This ... entails ... an assemblage of other fragments.... Becomings ... are ... something inherently unstable and changing. It is not a question of being (animal) ... but of moving, changing, being swept beyond the one singular position into a multiplicity of flows or into what Deleuze and Guattari have described as 'a thousand tiny sexes' ... to proliferate connections, to intensify.
>
> (ibid: 80)

Read in this way, perhaps the unlikely costume, setting and filming of the tango in *Reservaat* begin to make some sense, via Grosz's reading of Deleuze. The myriad of flows and proliferation of connections of a thousand tiny sexes, suggests the polymorphous perversity of a refigured productive desire no longer based on lack.

Spanton's performance in *Homeward Bound* might also be read in terms of a kind of becoming or transformation, since elements of her performance are most certainly 'inherently unstable and changing' (ibid: 80). This is evident in her deliberate performance of 'sailor boy' gestures and postures sometimes subtley hinted at and sometimes baldly stated as in 'Heave away me Johnny' when she leans on her upright lilo as if waiting for and watching the girls go by on a street corner (Plate 5). Spanton's androgynous looks combined with her costume, choreography and the words of the sea shanties, that refer to sailors' lost loves or to sailor boys which might also be the sailors' loves, open up gaps between embodiment and representation allowing for an eruption of a range of possible meanings. As the literary theorist Alice Parker says of Québécoise Nicole Brossard's lesbian writing '[she] meditates on gender, and how to tease out what is hidden in language' (Parker in Kaufmann, 1989: 229). Spanton too is playing with gender in her performance and teasing out what is hidden in the language of the sea songs such as 'Lovely Nancy'. The lyrics of the song overtly refer to a woman but given Spanton's undulating, rope climbing, serpentine body movements that accompany it, covert references to 'Nancy boys' are suggested. As Irigaray says: 'why only one song, one speech, one text at a time?' (1985b: 209). Spanton's display of nautical naughtiness erupts with thoughts, sensations and hesitations, giving it at times a coy, boyish appeal. Linking Deleuze's notion of 'becoming-imperceptible' with Irigaray's notions of the 'sensible transcendental', the philosophical theorist Tamsin Lorraine argues, 'becoming-imperceptible...involves bringing faculties to the limit of communication with one another... fragmenting the coherent subject and disabling the convergence of the faculties upon an object of common sense experience'. She continues this 'requires transformation of oneself as well as transformation of one's understanding of the world...no identity can remain fixed... becoming imperceptible involves challenging conventional body boundaries...and unsettling a coherent sense of self'(1999: 188–9). Spanton is transforming herself and the world in her performance, she is 'fragmenting the coherent subject and disabling the convergence of faculties on an object of commonsense experience' and as a result of her performance 'no identity can remain fixed', she 'unsettles a coherent sense of self'. Throughout her performance Spanton, partly because of the plays on the words of the songs, but also because of her ludic, polysemic choreography, continually shifts between subject and object. She is at times the object of our gaze, but then also the subject of the sea shanty. She is playing in the spaces in between in the same sense that Deleuze describes the operations of the word 'and'. He says, 'and is neither one thing nor the other, it's always in-between, between two things; it's the borderline...it's along this line of flight that things come to pass, becomings evolve, revolutions take shape' (1995: 45).

Spanton's subtly suggestive, fluid performance which moves between transitory identities is continually becoming in this sense; in a revolutionary sense. A multiplicity of meanings emerge from this intensely mobile, intertextual performance. 'There is a circuit of states that forms a mutual becoming, in the heart of a necessarily multiple or collective assemblage' (Deleuze and Guattari, 1986: 22).

Machinic assemblages in *Between/Outside* and *Reservaat*

For most of *Between/Outside,* Power and Spanton are seen walking up and down stairs usually passing each other, occasionally walking side by side, with hardly any signs of acknowledgement. Although the performers are near each other throughout, they are only seen to touch *four* times. The first touch occurs near the beginning when the dancers are sitting side by side on chairs. It seems significant because it occurs during a forty-five second long shot, noticeably longer than any previous shot, and quite long shots follow almost immediately, totalling one-and-a-half minutes in all, about a quarter of the video. Up to this point there is no indication that the women know each other, they simply keep passing on the stairs. They may be neighbours or complete strangers.

The sequence begins with Power entering from the front, turning and sitting next to Spanton to face camera. Neither woman acknowledges the other's presence. They continue looking straight ahead, as if they were sitting in a public place. There then begins a series of moves where each of them smooths their clothes underneath them and their skirts on their knees. In between this, Spanton sits on her hands, shifting her weight from side to side placing her hands under her, then removing them. Power places her hands first at her sides, then in her lap, crosses and uncrosses her leg, and looks slightly to the side away from Spanton. They then each look slightly towards each other, but their eyes never meet, they never acknowledge each other's presence. Throughout this, their inside forearms and elbows have been brushing against each other, almost imperceptibly. They appear rather like two people, possibly on a first date, gingerly seeing if chance touches might become something more. The performers' relationship is intensified by repetition within the choreography (each performer adjusts her position about eleven times in the course of one-and-a-half minutes), and by the mounting tension of these moments in and out of contact. The touches and the space between the women increasingly become centres of attention. This initially mundane contact starts to take on the character of a kind of foreplay or mating ritual. Two short shots between the three longer shots of the women sitting, also suggest contact. The women's legs and feet only are seen crossing a stair as if passing on a narrow ledge. The dancers must be very close, possibly holding each other in order to keep their balance. As Power (1999) says of *Between/Outside*: 'Video can suggest things by what it leaves out.'

In this excerpt when the dancers' arms are very close, the space between them increasingly becomes animated by choreographic repetition and by the duration of the phrase. The contrast with the lack of contact between the dancers in the rest of the piece renders these moments of touching particularly special. There is something about the ordinariness and drabness of the two women's appearance and situation in the impersonal, dull and empty stairwell that enhances the potential warmth and desire inherent in these unexpected moments of contact. It is often impossible to distinguish toucher from touched. This blurring of boundaries between the two results in an interconnectivity which is continually changing, creating a kind of machinic assemblage in a fluid state of becoming and transformation. Grosz, when discussing refigured lesbian desire, argues, 'the ways in which (fragments of) bodies come together . . . produce what Deleuze has called a machine . . . or assemblage' (1994b: 120). She explains that these interactions and linkages can be seen as both inside and outside (ibid: 116). The spatiality of the touches in this phrase is a continuous link from inside the subject to outside and from outside to inside. This blurs and decenters the boundaries of the two selves touching, eluding any kind of binary opposition. Precisely because it is impossible to distinguish toucher from touched and the image is of an assemblage, it is impossible to distinguish any clear unidirectional dynamic of desire between a self and an other marked as different. The desire is mutual and for the same. As Deleuze and Guattari state, 'the notion of behaviour proves inadequate' because it is 'too linear, in comparison with . . . the assemblage' (1988: 333). Initially the repetition in *Between/Outside* appears innocent, but as it increases it becomes apparent that this is an unfaithful repetition of performative acts in Butler's terms.[3] These women are not simply adjusting their clothing and sitting positions. There is much more at stake. This performance is clearly staged and constructed revealing the performativity and the discursively constructed nature of gender and sexuality.

Characteristics of Irigaray's 'two lips' imagery also apply. For example, she has stated: there is no 'possibility of distinguishing what is touching from what is touched' (1985b: 26). She writes of 'exchanges without identifiable terms, without accounts, without end' (ibid: 197) and of 'moving and remaining open to the other' (quoted in Whitford, 1991a: 161). In the apparently chance encounters of the two arms in *Between/Outside* it is often impossible to distinguish touched from toucher, and these exchanges do seem without terms, accounts or end. The two performers' movements also appear to remain open to each other not only in this encounter, but in others, such as the different passings, meetings and partings on the stairs.

About a minute after the 'sitting section' described above, both women are seen in profile in a close-up shot, Power is helping Spanton on with her coat. Underneath her red imitation fur zip-up jacket, Spanton is wearing only a black silk slip. Both women turn, face the camera and, standing side by side, look down to do up their coats. After zipping hers up, Spanton turns to watch Power, this is the first indication of acknowledgement. However, her look does not seem to be a significant new development in their relationship, it rather seems like a known repeated ritual that has occurred before. Furthermore Spanton's apparel, Power's help with her coat, and the fact that they are *both* putting on coats, suggest that not only do these women know each other, but maybe their relationship is quite intimate. The next few minutes show them simply passing on the stairs repeatedly, without eye contact. Again an obsessive ritual seems to be being acted out, but despite a small hint of acknowledgement in Spanton's look, there is no further sign that the couple know each other. They give nothing away, leaving viewers to ponder the nature of their relationship.

The next moment of contact occurs several minutes later when Power is seen helping Spanton on with her coat again. This time, after Spanton puts her arm in the sleeve, Power turns her to face her and zips her coat up for her whilst Spanton looks on. They briefly look into each other's eyes, before the shot changes. When the coat is being put on, the spatiality between the two women is filled by the coat that connects them and by the mundane everyday activity of putting on clothes. Grosz characterises machinic alliances as both relating through someone to something else and relating through something to someone, and claims that these connections form 'an intensity, an investment of libido' and this is what it means 'to see desire and sexuality as productive' (1994a: 79). The second time the coat is put on, Power does not leave Spanton to zip her own coat up, she turns her round to face her and does the zip up for her. The nature of the machinic alliance and spatiality bound up with the action has changed, transforming the assemblage and indicating the productive nature of desire. These machinic alliances suggest a 'different way of understanding the body in its connections with other bodies ... it is understood ... in terms of ... the linkages it establishes, the transformations and becomings it undergoes' (Grosz, 1994b: 164–5) and the blurring of its boundaries.

In *Between/Outside* there are several shots of legs only intertwining as they cross on a single stair. These shots of the dancers' legs, filmed in close-up, apart from the rest of the body can be compared with similar shots in *Reservaat*, glimpsed at the beginning and reiterated later. Towards the end of *Reservaat*, there is a view of feet and legs only. One dancer's foot kicks to the side behind the leg of her partner (performing a *gancho* or 'hook' – a kick performed between the partner's legs) to initiate a long running section. Alternate shots of legs and then bodies follow, the camera moving quite fast to keep up with the dancers. In the next shot, moving more slowly, the dancers step forward, executing a full turn around one of them performing a *giro* (a circular step), before starting to run again. Legs only are seen, they run sideways mirroring each other, ending with a jump.

The shots of intertwining legs in both dances can be described as 'assemblages of fragments', 'series of flows and breaks of varying speeds and intensities', 'moving, changing, being swept beyond the one singular position into a multiplicity of flows' – all images in Grosz's account of the Deleuzian notion of 'becoming animal' (1994a: 80). I suggest they could be extended to apply to any kind of 'becoming'. The difficulty of distinguishing one performer from another and imagining the relation of the rest of their bodies is an instance of blurred boundaries and binaries. In both dances the intertwined legs can be said to transform the space in which they move such that inside and outside are continuous and indecipherable. The spatiality from the inside of both subjects overflows and merges into a machinic assemblage animating the space between. As Deleuze and Guattari state, 'machinic assemblages are simultaneously located at the intersection of the contents and the expression ... they rotate in all directions like beacons' (1988: 73). In *Reservaat* at times the legs move so fast that they become blurred in the filming. The space around them is bound up in the becoming linkage. It too becomes sexualized.

In *Reservaat* the dancers' legs and feet are also seen touching themselves and touching those of their partner. These are highly sexualized rituals in the tango, based on traditional heterosexual notions of desire as lack. A foot or leg often penetrates the space between the other dancer's legs suggesting the sexual act. However, because the legs in *Reservaat* are filmed in isolation from the rest of the body it is difficult to see whose legs are whose, the conventional heterosexual signals are less apparent. The spatial dynamic is not about the binaries of presence and absence or activity and passivity, signalled when the desire of a subject is for an object marked as different. It is rather about a more productive desire for the same. In *Between/Outside*, where it is easier to distinguish between the dancers' legs because of their different shoes, the moves made by each are the same, blurring the binaries of active/passive and presence/absence. In both pieces, this intertwined machinic assemblage, as Grosz argues, provides 'a way of problematizing and rethinking the relations between the inside and the outside of the subject, its psychical interior and its corporeal exterior'(1994b: xii). The value of machinic assemblages is that they recognize, fit and reveal the complexities and multiplicities of desire seen as productive. In this sense these very different spatial linkages in the dances and the ideas of desire they suggest, whether they are concerned with arms almost imperceptibly touching, coats being put on, or legs intertwining, are all in Deleuze and Guattari's terms 'syntheses of heterogeneities'. As they state, 'these heterogeneities are matters of *expression* . . . their synthesis . . . forms a . . . machinic "statement" or "enunciation". The varying relations into which a . . . gesture, movement or position enters . . . form so many different machinic enunciations' (their emphasis) (1988: 330–1).

Polymorphous perversity and multiplicities in *Virginia Minx at Play*

A chaotic and overtly erotic geometry of multiple productive desire emerges in *Virginia Minx at Play*, as Claid plays with relationships between herself as a solo dancer, with the singer/musician, Heather Joyce, the spectators, and with herself. As Irigaray argues: 'these movements cannot be described as the passage from a beginning to an end. These rivers flow into no single definitive sea. These streams are without fixed banks, this body without fixed boundaries. This unceasing mobility . . . remains very strange to anyone claiming to stand on solid ground' (1985b: 215).

Claid appropriates the vocabulary and style of the distinctly macho and heterosexual Latin American dance for her own corrupt lesbian use. A tall imposing figure, in her close-fitting, high-waisted trousers and pink sequined waistcoat, Claid strides about the stage bathed in a deep pink light. She plays with the Latin vocabulary of dances such as the tango and flamenco, spinning fast, kicking high, arching her back and striking erect postures where everything in her body seems to be sucked up into her svelte torso. These dynamic dance phrases often end with seductive sideways glances at the audience from under the brim of Claid's fedora hat. She continually teases us grinning knowingly, winking and often reaching her upturned palm out to us as if offering herself, but then with an abrupt slap of her wrist she quickly withdraws, denying us our pleasure.

Claid then directs her attention to her accompanist, Joyce. She approaches Joyce, sitting at the side of the stage, from behind, boldly caresses her breasts, lies across her lap, pulls her up, turns her round and lowers her in a typical Latin dance embrace. When Claid returns her to standing, she faces her and slowly traces a line from the top of her forehead down the centre of her face and neck, between her breasts, over her stomach to her genitals, whereupon Joyce rushes back to the microphone to continue her song, leaving Claid with an empty outstretched hand. Spinning back to centre stage, her spangled, fringed waistcoat catching the bright lights, Claid goes down to the floor and reclines seductively on one hip looking at the audience. She then lowers herself onto her back, removes her hat and places it over her crotch, and whilst raising it, she fans the area between her legs as if to cool the heat. After one or two more sorties to Joyce, when Claid kisses her on the neck and strokes her, and several more plays with the spectators, the performance ends with a flourish as Joyce hits the high concluding note of the song and Claid sits down on the chair throwing a final sideways glance to the crowd. Here Claid is playing with traditional notions of desire based on lack. The singer is clearly the object of Claid's desire, evident in the ways in which Claid looks at her, looks at the audience knowingly, and then approaches her from a distance and touches her. Here, unlike the touches described between Power and Spanton in *Between/Outside,* there is no confusion as to who is touching whom, the touches are one-way. In this interlude Claid appears to be parodying traditional heterosexual desire based on lack.

However, in her performance as a whole, like Spanton in *Homeward Bound,* Claid's androgyny continually confuses as she plays recklessly with masculine and feminine codes. She deftly weaves her way through subject and object positions often occupying both at the same time as the chaotic geometry of desire criss-crosses the performance space. Claid's performance demonstrates many of the layers of deviousness possible when gender and sexuality are freed up in lesbian performance and let out to play. Claid's body exudes polysemic possibilities of a myriad erogenous zones from fingers to thighs, arms to mouths, breasts to torso and crotch, concurring with Irigaray's statement that: *'woman has sex organs more or less everywhere'*. She finds pleasure almost anywhere' (1985b: 28). Focuses of attention range continuously not only across Claid's own body but also over Joyce's body and back and forth between Claid and the audience, both disrupting and fracturing a voyeuristic, objectifying, masculine gaze, while also at times playing with it. For most of her performance Claid reveals 'female sexuality as fluid, diffuse, polymorphously perverse, mobile and unbounded' (Wilton, 1995: 155).

In-between spaces – spatial locations and private/public boundaries

The different spatialization of desire in the four dances examined here has been explored in, and created, in-between spaces of various kinds. In a literal sense the spaces between bodies of dancers have been seen at times to be animated rendering them special. Spaces between different body parts have also at times been animated in a similar way. The spaces between performers and spectators, live or virtual in the case of the dance videos, have also been importantly played with, particularly in the two live solos. In the two dance videos the added dimension of location provides another kind of in-between space. This blurs binary oppositions, creating further ambiguities which enhance and underline those inherent in the polymorphously perverse desiring bodies, providing particular contexts for reading desire as productive and spatialized differently. The blurred boundaries inherent in the touchings and closeness of the two bodies in *Reservaat* and *Between/Outside* are reiterated in their locations. The country park of *Reservaat* and the stairwell of *Between/Outside* are ambiguously both interior and exterior spaces. The park is enclosed by a wall and gate providing an interior space for animals and vegetation, but it is also an outside and exterior space signalled by shots of trees and sky above. Power, the choreographer of *Between/Outside*, says of it, 'the [stairwell] location is crucial: a space that is neither an interior nor an exterior, but one that lies somewhere between' (1999). The titles of both pieces, *Reservaat* and *Between/Outside*, reinforce the importance of the space of their locations and *Between/Outside* makes a statement about the in-betweenness of the space, further emphasizing the blurred boundaries of interior and exterior.

The continual blurring of interior/exterior and self/other oppositions is extended in both pieces when private/public boundaries are also challenged. In *Reservaat*, a seemingly private relationship, the intensity of which is signalled throughout, by close-ups of touches and looks, is acted out in a public space of a nature reserve. In *Between/Outside* the private/public divide is also played with in subtle ways. When the two performers sit side by side adjusting their dress and sitting positions, these apparently public moments increasingly appear private. When the two women who have appeared strangers help each other on with coats over underwear, and one takes time to lift and smooth the hair of the other, these moments seem surprisingly private. However, they might also be considered public, because of the apparently casual way in which they are slipped in among shots of two women who seem complete strangers, passing on what appears to be a public thoroughfare. Private/public boundaries can also be seen to be blurred in *Homeward Bound* and *Virginia Minx at Play* when the performers play in the space in between them and their spectators. They at times play with intimacies normally regarded as private in their public performances. The blurring of these public/private binaries, like the mutual touches in the dances, merges inside and outside, interior and exterior such that there is a fluid interface for the differently spatialized productive kind of desire evident in the works.

Conclusion

In these four dances when the touching, meeting and parting of the women's bodies, which are key threads that run throughout, are seen non-dualistically in terms of their spatialization as machinic assemblages, sometimes taking on the characteristics of 'becoming animal' or 'becoming imperceptible' with resonances with Irigaray's imagery associated with two lips, it becomes possible to conceive desire differently. The possibility of seeing desire spatialized in terms of linkages, connectivities and transformations suggests that desire can be seen as productive rather than as lack. This is because the boundaries of binary oppositions of self/other, subject/object, activity/passivity and presence/absence are often blurred in the dances allowing for a different spatialization of desire, emphasized in various ways by the choreography, and in *Reservaat* and *Between/Outside,* also by the filming. As Grosz indicates at the end of her essay on refiguring lesbian desire,

> what I am putting forth here is . . . a way of looking at things and doing things with concepts and ideas in the same ways we do them with bodies and pleasures, a way . . . of flattening the hierarchical relations between ideas and things . . . of eliminating the privilege of the human over the animal . . . the male over the female, the

straight over the 'bent' – of making them ... interactive, rendering them productive and innovative, experimental and provocative.

(1994a: 81)

The four performances explored have replaced desire based on lack, where space is seen as empty, with an animated space of touchings and intensities, and with seductive, actual and metaphorical plays with a wealth of errogenous zones. In the process, subject/object and self/other have been revealed as fluidly interchangeable with the potential to dissolve constraining binaries and challenge fixed identities. As Tamsin Lorraine claims, 'by delineating various aspects of human existence in terms of strata and assemblages, Deleuze and Guattari [and I would add these dances] depict a subject with vastly expanded possibilities and connections to the world' (1999: 172).

In the next chapter the expanded possibilities of subjectivity for contemporary urban Asian women are explored, focusing on the dissolution of West/East and male/female binaries, through the creation of in-between spaces in the choreography and filming of three female dancers in three London office buildings.

6
Hybridity and Nomadic Subjectivity in Shobana Jeyasingh's *Duets with Automobiles*

Introduction

In *Duets with Automobiles* (1993) three female dancers (Jeyaverni Jeganathan, Savitha Shekhar and Vidya Thirunarayan), filmed inside and outside three London office buildings (The Ark, Hammersmith, Canary Wharf and Alban Gate, London Wall), perform a mixture of vocabulary from contemporary dance and the traditional, Indian classical dance form of Bharata Natyam. I argue that the choreography and filming of the juxtaposition and interaction of the three female dancers with the geographically situated architecture construct these spaces as 'in-between'. In the process West/East and male/female binaries are blurred suggesting the possibility of a rethought, contemporary, urban, female subjectivity.

The West/East and male/female binaries that are blurred are suggested initially in the film by the juxtaposition of the dancers with the buildings. The office buildings have connotations of power, money and the city as a financial centre. They are created by men who are at the heart of controlling flows of finance. They are valuable and powerful properties because of where they are and what goes on in them; the control of capital; Western capital. They are masculine public spaces; women have historically been excluded from their creation and from power over their operations. Shobana Jeyasingh's choreography in this potentially alienating environment successfully manages to transform these spaces and fulfill her intention of creating 'an icon of Indian womanhood ... appropriate to urban women in the 1990s' (Rubidge, 1995: 34).[1] Jeyasingh achieves this intention through her distinctive approach to the Bharata Natyam dance form which she has metaphorically described as 'making a bedroom' out of the 'awesome public building' of the classical language of Bharata Natyam (in *Aditi* Newsletter, 1997: 9). In her choreography Jeyasingh is continually negotiating the distance between the formality, authority and sense of history that the 'awesome public building' of Bharata Natyam suggests, and the personal, human 'intimacy' and homeliness of the 'bedroom'. These negotiations of in-between spaces are both symptomatic of and informed by Jeyasingh's hybridity as a diasporic artist and her nomadic subjectivity.

Jeyasingh was born in India and now lives and works in Britain. When discussing her work she often refers to the postcolonial theories of Homi Bhabha (Jeyasingh, 1995, 1996, 1997b) whose concept of 'hybridity' she finds 'very useful' (in Ingram, 1997: 12). It gets away from the polarization inherent in descriptions of her dances as East/West collaborations. Bhabha is concerned to resist 'the binary opposition of racial and cultural groups' (1994: 207). His notion of hybridity recognizes the fluidity and changing nature of concepts such as 'East' and 'West'. The in-between spaces that Jeyasingh is continually negotiating in her work are hybrid spaces of the sort that Bhabha theorizes. He characterizes them as 'new areas of negotiation of meaning and representation' (1991: 211) which involve 'inscription and intervention' and 'translating and transvaluing cultural differences' (ibid: 242, 252). These terms aptly describe the actual and conceptual negotiations that occur in the in-between spaces created in *Duets with Automobiles* which is why Bhabha's theories are explored in the reading that follows.

Jeyasingh has said of the office buildings in *Duets with Automobiles* – 'we wanted to humanize' them and she has referred to them as 'an imaginary homeland' (1997a). There are references to Salman Rushdie's *Imaginary Homelands* (1991) in the monograph accompanying Jeyasingh's *Romance with . . . footnotes* video (Holmstrom, 1995: 6), and Jeyasingh has used his title for a conference paper (1995). Rushdie's 'imaginary homelands' describe the 'fictions' that 'exiles', 'emigrants and expatriates' create because they are unable to replace what has been lost due to their 'physical alienation from India' (Rushdie, 1991: 10). For Jeyasingh the concept of 'home' is an important one, she has referred to it on more than one occasion (Jeyasingh, 1996, 1997a, 1997b). Following Rushdie, she sees 'home' as an 'invention' and she talks about the 'fluidity' and 'unfixed' nature of notions of home (1996, 1997b). She relates these ideas to her background – after her birth in India, she grew up in Sri Lanka and Malaysia before coming to Britain. Like the feminist theorist Rosi Braidotti, who was born in Italy, raised in Australia, educated in Paris and is now based in the Netherlands,

Jeyasingh has actually experienced unfixed homes, and these experiences conceptually inform her outlook and her work.

Braidotti, introducing her theory of nomadic subjectivity, has written 'nomadism consists not so much in being homeless, as in being capable of recreating your home everywhere' (1994: 16). It is in part because of this consonance with Jeyasingh's ideas that Braidotti's concept of nomadic subjectivity is explored here. There are elements of Braidotti's theory that align with Bhabha's notion of hybridity, such as her stress on 'transmobility' and the importance of notions of 'interconnectedness' and her 'figuration' of the nomad as a 'culturally differentiated understanding of the subject' (ibid: 2–5). For her, nomadic subjectivity is a 'multiple entity' that allows for 'the recognition of differences' (ibid: 36). Her philosophy also entails 'a move beyond . . . dualistic conceptual constraints' (ibid: 2). But first and foremost Braidotti's nomadic subjectivity is a 'female feminist subjectivity' that is an 'attempt to move away from the phallocentric vision of the subject' (ibid: 1). Braidotti's moves towards a contemporary female subjectivity of this kind make her work particularly pertinent for examining *Duets with Automobiles* which Jeyasingh has stated is 'about strong, urban, Indian women' (1997a).

Duets with Automobiles (14 mins), commissioned by the Arts Council and the BBC and broadcast in 1993, is directed by Terry Braun[2] with music by Orlando Gough. It is the only work of Jeyasingh's that has been created entirely for camera. The work of Jeyasingh's company, formed in 1988 following her seven-year solo career in Bharata Natyam, 'questions inherited notions of what Indian dance is and what its possibilities are, stressing its belief that Indian dance is open to personal, contemporary and innovative use' (Rubidge, 1993). Although Jeyasingh's work is rooted in Bharata Natyam vocabulary and style, she develops it in various ways evident in *Duets with Automobiles*, such as her extension of the solo form to group choreography and her use of predominantly contemporary music.

Hybrid spaces between East and West

Jeyasingh has stated 'it's not a matter of choosing between this or that, we are already in a situation where the interconnections are so complex that we have enough work to keep track of that' (*Romance with . . . footnotes* video, 1994) and she writes of 'a blurring of a simple East West divide' (1997b: 32). Through dismantling the oversimplified binaries of East and West in her work, Jeyasingh is concerned to reveal the complexities and interconnections that exist on the borderlines in the many spaces between.

Duets with Automobiles opens with a shot of St Paul's Cathedral, seen through a window frame with a woman, her back to camera, silhouetted in the foreground. St Paul's marks the city skyline out as London, historical London, it is a national monument, part of the British cultural heritage. After a few seconds the woman slowly half turns to camera, with her dark skin she looks Indian. She looks back at the view and then turns back while contemplating herself slowly running her hand and forearm along the window ledge. This sensitive contact with the building, suggesting closeness and intimacy, initially contrasts with the view of St Paul's in the distance. Placing the Indian woman in the frame might suggest connections with British Imperial history. On one level the identities of London and of the Indian woman are constructed by their juxtaposition and the role that 'race' and history play investing them with power. The grandeur of St Paul's architecture could be said to dominate the scene investing London and British culture with power *over* the Indian woman.

Read in this way the image reinforces the East/West binary, but other readings are possible. The shot changes to a full-length view of St Paul's through a glass wall this time, in a golden, possibly evening, light with an Indian dancer standing, back to camera, silhouetted against the cathedral. She runs the palm of her hand over its surface through the glass. St Paul's, seen in the sunset, could suggest the remnants of an imperial past and that the dancer is contemplating the future of a new hybrid existence in a city which is hers. Jeyasingh said of the juxtaposition of the Indian dancers with the buildings in *Duets with Automobiles* 'it's about making something public very personal' (1997a). The caress of the dancer does this. It blurs the boundaries and binaries that separate her and St Paul's, it brings them closer together. Jeyasingh clearly had this in mind when making the piece. Discussing these opening moments, she said 'whenever I look at St Paul's, for me, it's a very ambiguous icon, because I see St Paul's immediately connecting to the Taj Mahal and to Santa Sofia in Istanbul. I wanted to pitch the Indian dancer against the image, I wanted her to caress it softly' (ibid). She continued stating 'I didn't want a fight or bitterness, it's not a power struggle' rather she saw these opening moments as being in sympathy with Bhabha's resistance to the binary opposition of colonizer and colonized. She said 'those dialogues are gone, I don't see East and West. I am not the "other" any more' (ibid).

This opening sequence sets the tone for *Duets with Automobiles*. Having the Indian dancer caress St Paul's foregrounds the ambiguity and hybridity of this icon. The intimacy of the looks and gestures of the Indian dancer soften the historic formality of the recognizably British public building, making it 'homely' and personal, private rather than public. The filming and choreography bring the two initially distinct elements together rather than placing them in opposition, creating an in-between space of hybridity, '[a] cultural space for opening up new forms of identification that . . . confuse the continuity of historical temporalities, confound the ordering of cultural symbols, traumatize tradition' (Bhabha, 1994: 179).

Relations between space, power and difference are not simple and straightforward as these opening moments of *Duets with Automobiles* indicate. Towards the end of the film the complexities of hybrid in-between spaces, with 'new forms of identification' that 'confuse', 'confound' and 'traumatize' in Bhabha's terms, become particularly evident. When a dancer kneels down on the classically, geometric designs inlaid in a marble floor, and traces with her palm some of the diagonal and circular lines of the pattern, she appears, in an act of reverence, to be making connections with the precise curved and linear pathways she has just traced in her performance of classical Indian dance. The 'mapping' refers to a parallel historical journey whose geometrical traces remain in the cultural products of dance, architecture and design. However, Jeyasingh, is also making a statement about the contemporary fluidity and movement of cultures and borders that currently allow such comparisons to be made. Talking of her work she has said she is concerned with 'the changing borders raging all around', the 'dynamism of journeys', and 'a pattern of belonging that is multi-dimensional' (Jeyasingh, 1995: 191–3). About her piece *Making of Maps* (1992/3), Jeyasingh said that it started as 'a process of inventing my own heritage'. One of the images that informed the piece was of someone playing music on a radio, twiddling the knob and sampling music from different countries. Jeyasingh wanted to express some of this 'amazing accessibility and openness of the universe that was there for the taking' (ibid: 193). Her mapped statements in *Duets with Automobiles* about traced connections between 'race', space and power, have resonances with the positioning, controlling and colonizing characteristics of real maps, inherited from history and demonstrated in dance, architecture and design (see Chapters 2 and 10). But they also challenge notions of identities and heritages fixed by borders by pointing to the fluidity of contemporary urban existence. This extract from *Duets with Automobiles* is a rich illustration of Bhabha's statement that,

> what is theoretically innovative, and politically crucial is the need to think beyond narratives of originary and initial subjectivities and to focus on those moments ... that are produced in the articulation of cultural differences. These 'in-between' spaces provide the terrain for elaborating strategies of selfhood ... that initiate new signs of identity.
>
> (1994: 1–2)

Bhabha's emphasis on the importance and value of language and culture in his work, that 'there is no knowledge ... outside representation' (1994: 23), is also illustrated in this extract by the hybridity evident in the dialectical articulation of the Bharata Natyam dancer's performance and the pattern on the marble floor. Here two cultural manifestations, which are in themselves hybrid, by their dialogue 'initiate a new sign of identity'. In this sense the choreography and filming reveal possibilities for reclaiming the space and imbuing it with new meanings. As Sanjoy Roy says of Jeyasingh's work, 'by getting "under the skin" of cultural boundaries, by loosening the links between race, place and culture, her work can speak to the experience of diaspora' (Roy, 1997: 83).

Female solidarity and nomadic subjectivity

In *Duets with Automobiles* Jeyasingh explores the multiple and shifting nature of questions of identity, specifically of the perspectives of her dancers, and their existence as contemporary, urban, Indian women. Several minutes into the piece, there is a shot of a dancer standing next to a circular white stone pillar curving her arm above her and placing her palm on the stone. This intimate gesture transforms the Bharata Natyam style of the movement. Instead of the emphasis being outward, on the frontal display of the solo dancing body as in traditional Indian dance, here the focus is inward, the body is turned sideways and the palm touching the stone connects the dancer to the modern London building (see also Plate 6). As she retraces her circular pathway with her arm, lowering her hand and spreading her fingers in a *mudra*, another dancer takes hold of her wrist and pulls her towards her. The first dancer lets her weight be caught by the second who is facing her, by placing her hands on her shoulders. The first is then turned and she leans back towards her partner who takes her body weight again, before sending her back towards the pillar. Touching it she turns and slides her back and the sole of her foot down it. The contact with the pillar and the interchange of weight between dancers are illustrations of Jeyasingh's concern to 'humanize' the buildings through touch and intimacy. The hints of female solidarity suggested also have connections with Braidotti's project to 'evoke a vision of female feminist subjectivity' (1994: 1).

As the dancers complete their interchange of weight, they are joined by a third and all begin to sink into full plié, where, in the next shot, they are seen with hands together in front of their chests in Bharata Natyam style. They stare out confidently to camera in front of the distant London skyline viewed through a glass wall. They rotate their hands in unison to form clenched fists, which they sustain as they turn and lower themselves diagonally onto their right knees. The image conveys their strong female presence. The clenched fist – a *mudra* called *mushdi* – was an important motif for Jeyasingh throughout *Duets with Automobiles* expressing strength and determination (Jeyasingh, 1997a). Next a single dancer leans forward on her knee and bows her head to the ground, where she rolls over onto her back and is seen in close-up, lying on the floor. Much of her body is touching the floor. The proximity, emphasized by the almost intrusive close-up shot, suggests an intimacy between the dancer and the building. She seems 'at home', inwardly confident, affirming an identification as a contemporary woman. All three dancers are then seen from outside, through the glass, sitting, staring out confidently through the transparent wall. They look to the side in unison. The shot changes briefly to a view of the London traffic, locating the dancers clearly in a modern, metropolitan environment.

The interdependency of women has been identified as a central theme in Jeyasingh's work. A Company monograph states Jeyasingh's 'double consciousness' of the Indian classical tradition and of contemporary urban cultural concerns is articulated when 'she introduces a new . . . relationship between her dancers through . . . extensive use of touch and weight giving' (Rubidge, 1995: 38) and 'she deliberately subverts the image of goddess or submissive, coy female . . . prevalent in the classical tradition, substituting . . . images . . . which more clearly express the behaviours of the contemporary Indian woman' (ibid: 34). Certainly in the excerpts just described there are examples of the dancers' interdependency through giving and taking weight and confident looks to camera suggesting strong, contemporary women.

These features of Jeyasingh's choreographic style, which portray a particular kind of female subjectivity and solidarity have links with Braidotti's theory of a specifically female nomadic subjectivity concerned to construct 'new forms of interrelatedness' (Braidotti, 1994: 2, 5). Certainly Jeyasingh's use of weight taking and support between her dancers is a 'new form of interrelatedness' for them. Braidotti's project is explicitly a feminist one, she asserts 'my task is to attempt to redefine a transmobile, materialist theory of feminist subjectivity that is committed to working within the parameters of the postmodern predicament' (ibid: 2). She emphasizes interconnectedness and the 'ability to flow from one set of experiences to another' (ibid: 5).

Jeyasingh's quest to create a contemporary, urban identity for Indian womanhood is illuminated by ideas about nomadic subjectivity. Braidotti argues, 'the nomadic subject ... allows me to think through and move across established categories and levels of experience: blurring boundaries without burning bridges' (ibid: 4). She has reclaimed the idea of the nomad as a positive metaphor for someone who is, in Bhabha's terms 'unhomed' rather than homeless, and is able, by thinking through and moving across boundaries, to create 'homes' anywhere. Jeyasingh is continually thinking through and moving across 'established categories' of the Bharata Natyam dance style which represents a cultural history of traditions and conventions about what an Indian woman should be and do. As a nomadic subject Jeyasingh is forever blurring and creating new borders and boundaries and negotiating spaces between them to create new 'homes' but without 'burning bridges' to Bharata Natyam. Bhabha writes of these spaces in between as a 'third space' which 'displaces the histories that constitute it' (1991: 211). His statement that this space 'sets up new structures of authority' and 'new political initiatives' that it is 'a new area of negotiation of meaning and representation' (ibid), aptly describes Jeyasingh's negotiations with Bharata Natyam.

Dancers and buildings and the spaces in between

In *Duets with Automobiles* an actual and metaphorical third space exists between the dancers' bodies and the architecture. This interface provides room where identities can be discovered, forged and played with. For the diasporic artist however,

> the exploration of this interstitial space can be seen as a series of creative conflicts: between a global, 'post-modern' critical hegemony,

and the critical codes of a particular history and tradition; between carving out a territory in the mainstream and making marginalisation itself a creative condition; between a shifting relationship with the host country (home) and a changing relationship with a notional 'back home'.
(Holmstrom in *Romance . . . with footnotes* monograph, 1995: 6)

These complexities are all apparent in *Duets with Automobiles* in the relationships between dancers' bodies and the architecture. The dancers' bodily shapes containing geometrical patterns, lines and curves, often within a circle in Bharata Natyam, are carefully placed inside or alongside circular domes, promontories and pillars, or next to the lines of window frames and walls. But in order to 'make a bedroom out of a public building' Jeyasingh goes beyond formal, aesthetic concerns and extends the Bharata Natyam language in a personal, emotional direction. The buildings become intimate spaces of touches, caresses, embraces, listening and sleep. A recurring motif is the downward movement of the palm of the hand usually on glass walls, but it also caresses stone pillars and wooden and marble floors of the office buildings. It acts as a kind of leitmotif that, together with the rest of the choreography and filming, marks the dancers' presence in an almost ritualistic fashion in these city spaces endowed with power associated with capital. The three dancers build an intimate relationship with the office buildings as, alongside traditional Bharata Natyam phrases of dance, they also roll, lie, kneel and place their heads on the floor as if listening or asleep, run, hug and lean on rounded stone pillars. These intimate points of contact of bodies and surfaces of the city buildings are emphasized by close-up shots, sounds, such as those of bare feet slapping out Bharata Natyam rhythms on wood and marble floors, choreographic repetition and other visual devices such as silhouettes. By their actions, through their close contact with the floors, pillars, windows and walls of these city spaces, the dancers transform the office buildings and turn them into intimate homely places. The film editing also contributes to this transformation. Shots change from one building to another seamlessly such that the three offices merge into one. Jeyasingh said 'we didn't actually want to separate the buildings, but to make them like one building, it was in some ways an imaginary homeland' (1997a).

There is a sense in which the presence, filming and choreography of the Indian classical dancers in the London office buildings – in these centres of capital and power – is a bold invasion of city space, which is in turn transformed by what occurs in it. One of the ways in which the space is transformed is through a particular bodily relationship with the architecture when the dancers repeatedly caress and hug pillars. This illustrates the complexity of the 'creative conflicts' between home and 'back home'. On one level this physical embrace of the modern urban building appears to be suggesting a metaphorical embrace of contemporary urban life. There is a sense in which intimate gestures such as these show the dancers making themselves at home in these empty offices, inhabiting them, humanizing them and making imaginary homelands out of them. However, Jeyasingh has indicated that she also had in mind the image of a *yakshi*; a female tree spirit often seen in classical Indian architecture touching a pillar or the building with a part of her body, usually her feet. This 'young fecund sort of female often with very big breasts' would be standing, carved into pillars of buildings symbolizing a source of energy since 'she's the creative principle making the tree or pillar come to life' (Jeyasingh, 1997a). Throughout *Duets with Automobiles* subtle references are made to this tree spirit through dancers sliding their feet down or hugging pillars. Bound up in these gestures are ideas about female energy or strength from the *yakshi* that relate to Braidotti's notions of female subjectivity, references to classical Indian architecture 'back home' and the transformation and translation of the gesture by its reinscription on the pillars of a contemporary London building. The gesture could be said to transform the building by making a metaphorical 'home' or 'bedroom' out of it at the same time as breathing energy into it or humanizing it through reference to *yakshi*. The dialogues and interactions between these different cultural ideas and images illustrate the complexities of the hybridity and nomadic subjectivity being explored.

Mutual construction of bodies and spaces

There are many ways in which new identities for the dancers and the buildings are suggested in *Duets with Automobiles*, since they mutually construct each other. Parallels can be drawn with the mutual inscriptions of bodies and cities that occur in *Muurwerk* and *Step in Time Girls* discussed in Chapter 3. In all three works the dancers transform the city spaces making them theirs, but the ways in which the choreography and filming achieve this are very different, as are the dancing bodies and city spaces involved.[3] In *Duets with Automobiles* there is a series of shots of the dancers seen from outside the buildings through glass walls. This is followed by an interior view of two dancers, one in the foreground and one in the background performing a *natya arambe* – literally an opening or beginning dance – against the backdrop of the city skyline. Next, the glass walls with reflections of traffic traversing them are seen from outside and three dancers are just about visible in the top right corner of the frame performing a plié, a tilt and a lunge in unison in Bharata Natyam style. This series of pictures of the dancers alternating from inside to outside the building has presented different views of them in relation to the architecture. Actual movement back and forth between different imagined spaces or notions of 'home' could be suggested; from dancers inscribed on the urban landscape, when viewed from inside the building, and the London traffic inscribed on the dancers, when viewed from outside. The buildings and the dancers mutually inscribe and construct each other and illustrate Bhabha's ideas of 'inscription and articulation of culture's *hybridity'* (1994: 38–9). These constructed spaces are 'racialized' in the process. The Indianness of the dancers, evident in their colour, costume and Bharata Natyam choreography, is inscribed on the city skyline when they are viewed from inside. When they are viewed from outside, with the glass edifice of the building foregrounded, the dancers, hardly visible, become overlaid with the reflections of London traffic; their Indianness is not nearly as apparent as it merges into the kaleidoscope of metropolitan images. They become part of the modern urban environment.

Perhaps one of the most memorable moments in *Duets with Automobiles* in terms of the ways in which the choreography reinvests the city spaces with a new kind of power is when a dancer travels forwards towards the camera framed by the walls and ceiling of a long corridor, emphasizing a sense of perspective as it recedes into the far distance. As she advances down the corridor slapping out Bharata Natyam rhythms with her feet on the marble floor, she is joined by first one and then the other of the three dancers. They move forward in unison, the sounds of their feet mingling with Orlando Gough's collage of rhythmic voices in the accompanying music. Their advance is emphasized by forceful arm gestures thrust directly towards the camera, sometimes led with a clenched fist. The three dancers complete their powerful surge forward exiting to camera. The impact of the phrase is further enhanced when it is repeated a little later in the piece, with two slight changes. The first is that the sequence is preceded by a long shot of the corridor further emphasizing the grand proportions of the neoclassical concourse against which the dancer initially is hardly visible. The second change is that this time the dancers perform unaccompanied, allowing the sounds of their feet on the floor to be heard uninterrupted, reinforcing the impetus of their advancing steps. The forward approach of the dancers seems relentless; it leaves an impression of female potency and strength that transforms this previously male-dominated centre of capital. The choreography and filming invest the city spaces and the bodies of the dancers with power and the possibilities of new meanings and new identities. The spaces become particular places and the bodies, empowered subjects, where the conflicts and contradictions of gender and 'race' are evident through the dancers' range of interactions with the architecture.

Nomadic subjects in cities

At the end of *Duets with Automobiles* a colonnade outside a building is shown and the three dancers appear in everyday dress, sitting on the ground in the sun, leaning against a large pillar; they are *outside* the building and they are not dancing. There is no music, for the first time natural sounds are heard. The building has not been seen before, there are no familiar marks of identification. The three Indian women are chatting, they look up as someone passes by and then continue their conversation. They could be in Bombay (one shields her eyes from the sun), or Birmingham, Manchester or Madras. London is no longer the only possibility. The anonymity of the setting might suggest that these women are 'unhomed' or 'unhomely', in Bhabha's terms. As Jeyasingh claims,

> the culture of the diaspora challenges inherited ideas of home as something defined by geography. More and more 'home' is becoming a radical and dynamic intervention that is more about the future than the past, the sum of new journeys rather than the station that one has left behind.
>
> (1997b: 32)

What Jeyasingh said about the end of *Making of Maps* (1993) might also apply here. Talking of the role of urban sounds in the piece, she said

> For me it was very important to bring back the sounds of the city ... it was the only way to end the dance. The whole dance was ... a questioning journey that an Indian dancer in Britain was asking about where she belonged, so it was fit and proper that the dance actually started with the street sounds because that is where we live ... in these big cities, at the same time ... we carry the very Indian

part of us... but... we don't make this journey to India and stay there, we've got to come back and find a resolution, find a peace within the situation which we find ourselves in now. So it's very proper that the dance comes back to the everyday sounds of London, or Birmingham... or wherever we live, and that we find a new way of dancing with the music of the everyday.

(*Making of Maps* video, 1993.)

Ending *Duets with Automobiles* with 'the everyday' reminds audiences that these dancers are also contemporary women who live in cities. They have the inward assurance as nomadic subjects to inhabit the in-between spaces of hybridity.

Conclusion

Exploring Bhabha's theory of hybridity and Braidotti's theory of nomadic subjectivity alongside *Duets with Automobiles* has highlighted the ways in which the work explores in-between spaces and specific consequences of this exploration. Placing three Bharata Natyam dancers in offices which have derived power from controlling capital, and are hence at the heart of the city, seems at first sight possibly an extreme imposition. However the innovative choreography and filming which are sympathetic and sensitive to the aesthetic concerns of the architecture and the dance vocabulary reveal the potential for redefining the dancing bodies and the city spaces. The physical spaces between the dancers and the architecture are highlighted in the choreography through various kinds of contact such as caresses, embraces, leaning, rolling on, – specific touches and looks. These means of contact are carefully chosen such that they both extend and develop traditional classical Bharata Natyam vocabulary with contemporary inflexions. As a result differences between past and present, classical and contemporary, East and West, and female and male in terms of dance and space are blurred. It is in this sense that Jeyasingh's choreography subverts and intervenes by conceptually dismantling binaries and challenging hierarchies and entrenched cultural values. The in-between spaces of *Duets with Automobiles* are also emphasized in the filming through close-ups, framing, editing and filming through and against glass windows and walls. Thus, the relationships between the dancers and this city space of capital are continually foregrounded and as a result, the cultural differences that exist are conceptually, 'reinscribed', 'translated' and 'transvalued', such that these spaces become 'new areas of negotiation, meaning and representation' in Bhabha's terms. In this sense, as Bhabha argues, such 'borderline work' 'demands an encounter with "newness"', 'it renews the past refiguring it as a contingent "in-between" space that innovates and interrupts the performance of the present' (1994: 7). The performers, through their dancing and the ways in which they are filmed, seem to enact a ritual in these city spaces invested with power. They subtly, yet firmly, assert their presence and in so doing, their new hybrid identities as contemporary, urban, Asian women are affirmed.

Braidotti's theory of nomadic subjectivity has parallels with Bhabha's theory in terms of its focus on dismantling and blurring the binaries and boundaries of, in Braidotti's terms, 'race, class, gender and sexual practice' (1994: 2). What her theory adds are particular focuses on the importance of transmobility and the value of a specific female solidarity associated with interconnectedness. As a result, reading *Duets with Automobiles* with Braidotti's theory in mind, highlights the ways in which the choreography and filming emphasize the mobility of the dancers in the city space both physically and conceptually. The dance is filmed inside and outside three different London office buildings but the film's editing makes it difficult to fathom which building is which. There is a sense of the dancers' mobility between and confidence in these spaces. Their nomadism is empowering and, at least partly as a result of their ability to move between these different spaces of capital and power, they appear to be 'at home' in them.

The other key contribution of Braidotti's theory is its inherently feminist focus. When it is explored alongside *Duets with Automobiles*, attention is drawn to the spaces between the dancers. These spaces can also be seen to be conceptual in-between spaces like those between the dancers and the architecture. Jeyasingh does not simply extend the traditional solo form of Bharata Natyam into a group form, she pays particular attention to the ways in which her three female dancers relate to each other. They support and lean on each other, lift and hold each other in ways that emphasize both closeness and strength. In her discussion of *Making of Maps* she writes of the ways in which she wanted to 'question ... the self sufficiency of the Bharata Natyam body conditioned by its history as a solo art form'. She suggests, 'it was not just a matter of introducing touch and with it the implication of human emotion and relationships, but also the acceptance of physical dependency and trust' (1997b: 32). Physical dependency and trust are evident between the dancers in *Duets with Automobiles* when they help each other off the floor, assist each other in jumping by lifting, and when they take each other's weight. In this sense, they illustrate Braidotti's stress on female solidarity within nomadic subjectivity which, according to her, constructs 'new forms of interrelatedness' and connectedness.

The new forms of interrelatedness in *Duets with Automobiles* occur in the in-between spaces created and foregrounded between dancers, and between dancers and buildings representing city spaces associated with Western, male power. In this sense, the dancers in *Duets with Automobiles* through their interconnectedness with each other and the buildings have enunciated a new kind of empowering, female subjectivity, which through its hybridity, also forges new relationships with the cultures of old and new homes. They are able to do this because of the ways in which the choreography and filming show them 'at home' in the city office spaces and at the same time making a 'bedroom' out of the traditions of the 'awesome public building' of Bharata Natyam.

In the next chapter actual and metaphorical in-between spaces created by constructions of cultural difference on both sides of the Atlantic are explored in two works focusing on issues of identity, power and difference. While experiences of displacement and subjection are seen to contribute to constructions of cultural difference, resistance is revealed as empowering, suggesting possibilities for fluid identities and subjectivities with space for celebrating difference.

1 Film still of Roxanne Huilmand in *Muurwerk* courtesy of argos international film distributors.

2 Photograph of Yolande Snaith in *Step in Time Girls* by Ross MacGibbon reproduced with his permission.

3 Photograph of *Out of the Windy Beach* by the author.

4 Photograph of Emilyn Claid in the 'Watch me Witch You' section from *Virginia Minx at Play* by Eleni Leoussi reproduced with her permission.

5 Photograph of Sarah Spanton in *Homeward Bound* by Michelle Atherton reproduced with her permission.

6 Still from *Duets with Automobiles* from the series 'Dance for the Camera' courtesy of Arts Council/BBC.

7 Photograph of *Ellis Island* by Bob Rosen reproduced with his permission.

8 Photograph of Trisha Brown in *If You Couldn't See Me* by Joanne Savio reproduced with her permission.

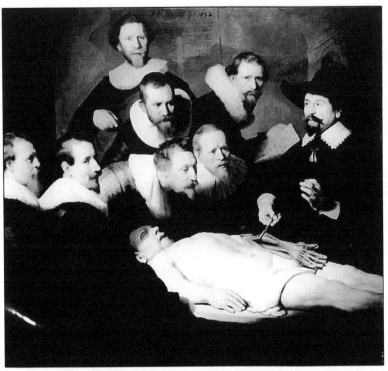

9 Photograph of *The Anatomy Lesson of Dr. Tulp* (1632) by Rembrandt courtesy of Mauritshuis, The Hague.

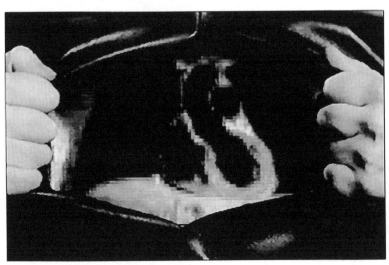

10 Still from *Joan* courtesy of MJW Productions.

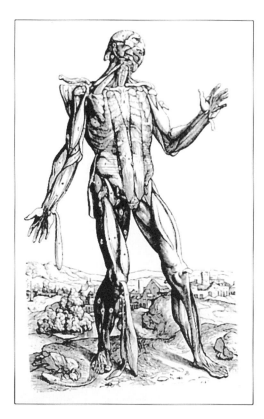

11 Photograph of Andreas Vesalius's illustration – The Fifth Plate of the Muscles from the Second Book of the *Fabrica* reproduced with permission of Dover Publications Inc.

12 Still from *Blind Faith* courtesy of Yolande Snaith Theatredance.

13 Photograph of Liz Aggiss in *Grotesque Dancer* reproduced with permission of Divas.

14 Photograph of CandoCo Dance Company in *Across Your Heart* by Chris Nash reproduced with his permission.

15 Photograph of *Rosas Danst Rosas* by Herman Sorgeloos reproduced with his permission.

16 Photograph of *Enemy in the Figure* by Dominik Mentzos reproduced with his permission.

7
Crossing the (black) Atlantic: Spatial and Temporal Displacements in Meredith Monk's *Ellis Island* and Jonzi D's *Aeroplane Man*

Introduction

Whereras the first two chapters in Part II focused on in-between spaces derived from actual spaces between dancers (in Chapter 5), and from diasporic spaces (in Chapter 6), a discursive in-between space provides a symbolic reference for the exploration of displacement in the context of the politics of migration in this chapter. *Ellis Island* (1981) and *Aeroplane Man* (1997) are both concerned with experiences of displacement in in-between spaces on actual and metaphorical journeys which are like rites of passage. This chapter examines how these two works explore issues of identity, difference and power, spatially and temporally in the course of these journeys.

The British postcolonialist Paul Gilroy's concept of the 'black Atlantic' illustrates the complexities of identity formation that arise out of various spatial and temporal displacements. The idea of the black Atlantic represents not only the enforced crossings of slaves from Africa but also return crossings of liberated African Americans in search of education and employment in Europe. It reveals the 'complex of difference and similarity that gave rise to the consciousness of diaspora inter-culture' (Gilroy, 1996: 20). *Ellis Island* and *Aeroplane Man* are both concerned with how the diasporic movements of peoples at different times 'muddy the waters' of identity formation. *Ellis Island*, a film created by American artist, Meredith Monk, shows immigrants, black and white, who have crossed the Atlantic being processed as they pass through the US immigration centre on Ellis Island in New York harbour in the early twentieth century (Plate 7). *Aeroplane Man*, a rap performed by British artist Jonzi D, tells of his semi-autobiographical travels back and forth across the Atlantic in search of his roots. These works, by revealing the roles which space and place play in identity formation and the construction of difference, expose some of the causes of displacement. In-between spaces formed by the construction of borders are key in both works. The immigrants on Ellis Island are displaced in a liminal, in-between space on the borders of the state they wish to enter. Jonzi D experiences the marginality of displacement in Britain and the in-betweenness of not belonging as he crosses borders on his travels.

The role of power in creating identities based on difference that result in feelings of displacement is exposed in both works. The French poststructuralist Michel Foucault's theories of disciplinary technologies of power when explored particularly alongside *Ellis Island* reveal the ways in which power creates 'others' through constructions of difference. Both works show how these discourses of power are perpetuated and how their effects, which importantly include resistance to power, construct and contribute to experiences of displacement, but also to alternative subjectivities where difference is valued.

I see *Ellis Island* and *Aeroplane Man* as complementary. Where *Ellis Island* focuses on institutional mechanisms which construct subjects of difference, *Aeroplane Man* explores the results of this construction in an individual's experience. They are also complementary in their spatial and temporal treatment of these issues revealing the interdependence of these two dimensions. Their exploration of in-between spaces draws on and points to histories of racial and colonial oppression while indicating their contemporary relevance. *Ellis Island* shows the immigration centre in operation at the turn of the century and as contemporary ruined buildings which tourists are shown around. In *Aeroplane Man* Jonzi D, drawing on history, travels to Africa and the Caribbean in search of his roots, but his performance reveals that these journeys are prompted by his *contemporary* experience of racism in Britain. Both works explore history and memory but 're-order them within patterns of meaning which belong to the present' (Betterton, 1996: 175).

The in-between spaces of *Ellis Island* and *Aeroplane Man* are in the margins, characterized by the African American feminist bell hooks as being 'both sites of repression and . . . resistance' (1990: 342). Mechanisms of institutional repression are portrayed in *Ellis Island*. In *Aeroplane Man* the repression of contemporary racism in Britain is depicted. Resistance is evident in affective elements in the works which exist outside the symbolic boundaries of the contained subject constructed by language. These affective elements are evident in dance and music performance in both works. A rare glimpse of the immigrants enjoying themselves waltzing in *Ellis Island* can be read as a moment of resistance. The politically charged protest theatre mode of rap performance used in *Aeroplane Man* is inherently resistant, particularly in the ways that it wittily plays with language. Both pieces illustrate Gilroy's (1995) claim of the potency of music and dance performance as signs of resistance in the history of repressed peoples.

Both works also identify the role of language in operations of power, subjectification and displacement. In *Ellis Island* language is revealed as a tool of normalization; through language lessons which the immigrants are given and through their labelling and documentation by that language. In *Aeroplane Man* Jonzi D is displaced by the language of others which he vividly describes through the distinctly provocative and empowering language of his rap.

After an introduction to the artists and works being considered, the constructed nature of identity is explored. It is argued that both works demonstrate the role that difference plays in the construction of identity; they show that identity has no fixed origins, that it is multiple and fragmented; and they illustrate how identity is a process of becoming, rather than a descriptor of a stable state of being. It is argued that *Ellis Island* and *Aeroplane Man* illustrate the important role place plays in identity construction. Next, using Foucault's theories, I argue that *Ellis Island,* in particular, reveals technologies of power that contribute to the identity construction process and result in displacement. The role of resistance as an effect of power is identified and the important role that language plays in the identity construction process is explored. I argue that both works demonstrate how spatial containment, brought about by the construction of borders which territorialize, contributes to identity construction and displacement, but also how the blurring of borders challenges containment revealing ways in which difference can be appreciated.

Meredith Monk and Jonzi D

Meredith Monk is a white Jewish American woman in her late 50s whose career spans almost forty years and Jonzi D is a black British male in his 20s. Although of different backgrounds, gender and generations, their works share certain key characteristics. Both are semi-autobiographical drawing on 'racial' and family histories, both use travel as an image to explore identity construction, and both are multi-media works blurring the boundaries between dance, film, theatre, music and text. These common features are significant in the explorations of displacement in *Ellis Island* and *Aeroplane Man*.

Meredith Monk began performing in New York in the early sixties and formed her own semi-communal performing arts company, The House, in 1968. Key features of Monk's work evident in *Ellis Island* are her use of pedestrian movement, ritual, archetypes, cinematic syntax and tableaux[1] to portray humanity and a sense of community. She has said that she wants to give audiences the opportunity 'to sense the fullness of their experience and the fullness of all the aspects of themselves' (in Zurbrugg, 1993: 98). She clearly values the differences evident in world cultures and there is a sense in which *Ellis Island* celebrates this. Her non-linear plays with time, including past and present in the same work, with her use of radical juxtaposition, give her work a dreamlike or surreal quality. Her interest in space and history has resulted in several site specific works and a focus on archaeology. Monk has said, 'when you create ... it's really a process of uncovering' (in Baker, 1984: 3). *Ellis Island* featured originally as a seven minute film in *Recent Ruins* (1979), one of a series of archaeology based pieces. It was extended to 30 minutes and filmed in 35mm colour and black and white in collaboration with filmmaker Bob Rosen and photographer Jerry Panzen.[2] Broadcast in West Germany, the USA and Britain, and released on video, it has won many awards. Filmed in the derelict, abandoned Ellis Island immigration centre, its black and white sections consist of a collage of images of the place, objects within it, and tableaux of immigrants posing for photographs, being measured, examined and labelled. There is hardly any 'realistic' sound, the accompaniment is a selection from Monk's own haunting vocal compositions. These historic images are interspersed with film shot in colour with naturalistic sounds of contemporary guided tours of the centre.

Jonzi D (actual name David Johns) started rapping and breaking[3] in 1984, graduated from the London Contemporary Dance School in 1992, and was appointed choreographer in residence at the associated Place Theatre in 1998. His background is in South London hip hop culture which he merges with his conventional dance training. He formed Lyrikal Fearta, a company of rappers, dancers, musicians and a DJ, to fuse abstract imagery and hip hop culture. He sees his work as 'the voice of the underclass in ... capitalist society' 'the very personal honest expression of how I see the world' (Jonzi D, interview on *Aeroplane Man* video, 1997). *Aeroplane Man* existed originally as a short solo within a full-length show, *The Requel* (1997). It has since been videoed as part of

The Place Spring Loaded 3 (1997) series and extended for a full evening performance, lasting approximately one hour and 45 minutes, at The Place for a week in April/May 1999. The seven minute video of *Aeroplane Man* consists of Jonzi D performing his rap solo in a single spotlight, centre stage. The rap relates his experience of racism as a British black and his consequent decision to leave in search of his roots. 'Call up Mr Aeroplane Man' is a repetitive refrain that Jonzi sings while jogging, he asks 'Mr Aeroplane Man' to jet him to Grenada, Jamaica, America and Africa. On each 'visit' Jonzi D performs the typical vernacular dance of the place and enacts encounters with people he meets. The piece ends with him jogging, asking 'Mr Aeroplane Man' to 'Keep on flying'. Throughout Jonzi D is cheered by the audience. In at least one live version of this (1997) production the piece opens with a film showing Jonzi D jogging through London heralding his arrival in the performance space (Leask, 1997: 38). In the extended version of *Aeroplane Man* Jonzi D is joined by a company who perform the dances and music of the places he visits. There is also an opening mime scene where Jonzi D, lying under a Union Jack covered duvet centre stage, has neatly folded piles of clothes representing figures of authority; a policeman, a judge and a teacher, placed around him. The piece ends with Jonzi D being stripped to his Union Jack underpants by the Africans he meets and then going through a ritualistic, transformatory healing or cleansing rite of passage with a traditional African female 'witch doctor' figure. This is partly symbolized by her dressing him in white trousers and top, but also by his angry exclamations about Britain which he gets out of his system, and by four other performers dressed in black lifting and moving Jonzi D through the air in a slow motion dream-like sequence. The piece closes with him repeating 'many manifestations of meaning' over and over until his final line: 'this brown frame has found his name', indicating the completion of his rite of passage and the beginnings of a new subjectivity, that is open rather than fixed – he doesn't state his 'name'. *Aeroplane Man* includes improvisatory elements, each version is different, although the basic structure of Jonzi D's journey remains the same.

The infectious style of this semi-improvized African format, valuing sponteneity and audience interactions typical of a community of shared values, is like *Aeroplane Man's* ending, appropriately open rather than fixed. Jonzi D has a photograph of Rennie Harris, the African American hip hop dancer/choreographer, who also employs Africanist improvization, on one of his programmes. The statement of African American dance theorist Brenda Dixon Gottschild that Harris's 'intention is to allow dance to be the connective tissue for bridging and embracing the seemingly contending opposites: black–white, male–female, self–Other' (1996: 159) could also apply to Jonzi D. Harris's 'qualities of openness and giving' which point towards 'a potentially multicultural world in which difference is celebrated' (ibid) seem to be an inspiration, and Harris's statement that 'I've decided to start a healing process that will enable me to face my deepest fears... by healing myself I'm healing my oppressors' (1994: 17) seems particularly pertinent to Jonzi D's *Aeroplane Man*.

The constructed nature of identity

At the centre of *Aeroplane Man* there is a frustrated search for lost origins that might inform identity, and in *Ellis Island* immigrants' identities and differences are foregrounded. As I indicated in the Introduction both sameness and difference lie at the heart of subjectivity, both these works demonstrate this fundamental instability and the complex constructed nature of identity, as a key constituent of subjectivity.

Identity constructed through difference

Jonzi D's search for his roots is a search for his identity, for who he is. He discovers a significant element of his identity in each place that he visits because of his *difference*, because he is *not* like the people he meets. A sense of self is defined in relation to an 'other' who is different. Jonzi D is made aware of his difference because he comes from Britain. For example, in Grenada he is called 'English boy'. Place is a significant factor here, 'notions of identity and alterity, of "us" and "them", are closely linked to the sense of place... to... notions of "here" and "there"'(Schick, 1999: 23).

Ellis Island graphically shows how identity is constructed with reference to the body as a marker of difference. At one point a hand draws a circle with a marker pen around the nose of one of the immigrants. This bodily feature has been highlighted as a sign of difference. When the hand then writes 'J' in a circle next to this woman's face, presumably standing for 'Jew', the sign of difference has been interpreted as a marker of 'race'. By presenting the labelling as 'racial' alongside other practices of the immigration centre, the film shows that it is not innocent. It is part of the mechanisms of power at work. The film situates this as part of a larger picture which shows that 'the "difference" of the post-colonial subject by which s/he can be "othered" is felt most directly ... in the way in which... superficial differences of the body... are read as indelible signs of the "natural" inferiority of their possessors' (Ashcroft *et al.*, 1995: 321). By this act, the body in *Ellis Island* becomes 'the inescapable, visible sign of . . . oppression and denigration' (ibid).

In *Ellis Island* groups of 'us' and 'them' are constructed through relationships to place. The immigration centre defines people's identity as either immigrants, or immigration officers. Difference is created by how the two groups are filmed, how they are dressed and what they do. One group is predominantly active and 'does things' to the others who are predominantly passive. The latter group is often seen posing for photographs. One group is more smartly dressed than the others, sometimes in uniform. This group exercises power over the others by its actions. The immigration officers are constructed as subjects, as 'I's, and the immigrants, as objects, 'not-I's or 'others', by these acts. By medically examining them, surveying, interviewing and teaching them, the officials are 'othering' the immigrants, giving them identity as 'others' through difference.

Fragmentary identity without origins

Each place Jonzi D visits in search of his roots in *Aeroplane Man* represents a fragment of his multi-faceted history and genealogy. No one place provides him with a sense of origin. There is no such thing as a single pure identity deriving from one source. Jonzi D is not accepted as having origins in Grenada, Jamaica, America or Africa, because he is English. For example, in America, in the words of his rap he is called an 'English nigger' which illustrates his fragmentary identity, and he is told to 'Get the fuck outa here.' Through his travels Jonzi D and his audience come to realize that, as Gilroy argues, 'no straight or unbroken line of descent . . . can establish plausible genealogical relations between current forms . . . and . . . fixed, identifiable . . . origins' and that 'the forbidding density of the processes of conquest, accommodation, mediation and interpenetration that . . . define colonial cultures . . . demands that we re-conceptualise the whole problematic of origins' (1995: 15). The density of these processes of mediation and accommodation gets in the way of and ultimately prevents Jonzi D's identification with the people and places he encounters on his travels.

One hint *Ellis Island* provides of the immigrants' possible origins is a citing of plates of food. Single shots of a plate of beans, of potatoes and of spaghetti are interspersed between shots of people posed as if for photographs. The singling out of these foods associated stereotypically with Mexico, Ireland and Italy fixes them as signs. Stereotypes work to fix – all Mexicans eat beans, all Irish eat potatoes. This kind of fixity is as unstable as the notion of origins. Eating beans does not *mean* being Mexican and 'being Mexican' is not a simple unified concept. Identity is a social construction which is fragmentary, complex, multifaceted and changeable. It is made up of many factors including nationality, 'race', class, gender and so on. Each person's identity is constituted through an interaction of these factors. Juxtaposing fixed images of plates of food with fixed images of people suggests that this is how the immigration authorities saw the immigrants. The portrayal of plates of food as oversimplistic markers reveals them as contingent constructions, and suggests that similar oversimplistic markers were used to construct the immigrants' identities, based on notions of fixed origins which, by juxtaposition with the food, are revealed as contingent.

The fragmentary and multiple characteristics of identity are repeatedly emphasized in *Ellis Island*. History and genealogy are referenced through archaeology. A wall is shown marked at different levels with dates '1890', '1920', '1954' and '1985' like an archaeological excavation. Juxtaposed with parallel views of the past and present immigration centre, accretions of time, geologically apparent in layers of earth, but also in people's lives and histories, are revealed. The film is suggesting that an individual's consciousness and unconscious identity, like the earth, are layered, forming multiple, fragmented subjects. By including archaeological metaphors alongside the tableaux of immigrants a kind of 'counter history' is revealed. This, as Grosz has argued, is 'uneven' and 'scattered', made up 'of interruptions, irruptions, outbreaks and containments' characteristic of subordinated social groups (1990: 78). The non-linear form of *Ellis Island* structured as a collage of images, mirrors the non-linear fragmented histories of the people it depicts. The *Ellis Island* immigrants are diasporic. They have undertaken between them multiple journeys and dispersed from different places or 'homes'. Diaspora is concerned with historical displacement and often 'home' is 'a place of no return' (Brah, 1996: 192), as is evident in the discussion of notions of 'home' in Chapter 6. On Ellis Island between 1892 and 1927 approximately 3000 rejected immigrants 'committed suicide ... rather than face deportation to their country of origin' (*Ellis Island* Publicity Flier, 1981). For them, home was a place of no return. In this context discourses of fixed origins and simple, unified identities deriving from a 'home' to which it is possible to return, are meaningless, the concept of diaspora critiques them (Brah, 1996: 197).

Identity as process

'Since identity is process, what we have is a field of discourses, matrices of meanings, narratives of self and others, and configurations of memories... Every enunciation of identity... in this field of identifications represents a reconstruction' (ibid: 247). *Aeroplane Man* and *Ellis Island* contain enunciations of identity that can be seen as reconstructions. Each time Jonzi D visits a foreign land in search of his roots, he engages with 'fields of discourses, matrices of meanings, narratives of self and others and configurations of memories'. He is going to those places because of what he 'knows' about them, because of what he has been told (stories about Grenada by his parents), because of what he has read (about Africa), because of what the media has shown him of America. He raps hopefully: 'Jet me to America land, on TV and music video we see enough black man, seems like they've got a plan'. What *Aeroplane Man* reveals is that the discourses, memories and narratives Jonzi D takes with him are part of complex ever changing configurations and matrices of meanings that constitute a field of identifications. Each of Jonzi D's encounters with different people and places is an enunciation of his identity that also proves to be a reconstruction through displacement. In each place Jonzi D's versatile body, energized by Caribbean, American hip hop or African musical sounds, slips easily into the vernacular dance style. His performance explores possible identities informed by the cultural histories of the places he visits. His encounters change and displace him. On another level Jonzi D's rap is a narrative which 'plays a central role in the constitution and preservation of identity. It is a carrier of meaning, the channel through which [Jonzi D]... tells himself and others the tale of his place in the world' (Schick, 1999: 21). It is an example of identity as an effect of narrative or discourse.

In *Ellis Island* identity in the making is accentuated by being revealed as a process of 'becoming rather than being'. In one scene a uniformed official asks an immigrant his name. No dialogue is heard, the official's lips mouth the word 'Name?' which appears on the screen. The back of the immigrant's head is seen moving, presumably he is saying his name. 'Ellessen Rahmsauer' appears on the screen, followed by a series of misspellings of the name until it becomes 'Elie Ram'. The absurdity of what is occurring is emphasized by the final two names that appear, 'Eli Sheep' and 'Eli Lamb'. Sheep are notoriously docile animals that 'follow' and they have a reputation for going 'astray'. These associations could be seen to apply to the Ellis Island immigrants. Their passivity renders them docile and their homeless status makes them appear 'lost'. The application of Foucault's theories of subjectification explored later shows how the disciplinary technologies revealed in *Ellis Island* produce 'docile bodies'. 'Ellessen Rahmsauer' has gone through several transformations to become 'Eli Lamb'. Becoming rather than being is evident in this incident where time and history are fast forwarded revealing genealogy as process. By escalating the process and including the ludicrous changes from 'Ram' to 'Sheep' to 'Lamb' the film underlines the lack of respect of the officials for the immigrants accentuating the power differential between them. This deconstruction makes the recording of immigrants' names seem like some bizarre word association game. The naming that occurs shows identity as a process of becoming in an authoritarian context which attempts to fix it.

Another example from *Ellis Island* illustrates how identity is constructed and the power operations that work to fix it. When the word SERB is written next to a man's face, this labelling classifies him. Classification is a process of fixing, involving selection and ranking, championed in the name of science. '[The] hegemonic project [of science] confidently stalked the world identifying . . . and classifying fauna, flora and peoples; asserting its "scientific neutrality" while marking hierarchies of "race", class, and gender' (Brah, 1996: 221). The key word here is *hierarchies*; classification of this kind shows the operations of power at work that underlie racism. By showing the labelling of an immigrant as SERB alongside other inscriptive processes, the film reveals the contribution of Ellis Island's immigration project to the broader hegemonic project, and exposes the racist identity forming inherent in it.

Aeroplane Man demonstrates that identity is not simply a fluid ever changing process. There are actual groundings in terms of the effects of identity that have to be faced. The opening mime scene in the extended version is evidence of this. The slow ritualistic placing of piles of neatly folded clothes by a white male pacing out a triangle in the semi-darkness around a presumably sleeping Jonzi D, under the Union Jack duvet, is an eerie beginning. In the half light spectators can gradually perceive the piles of clothes as a policeman's uniform, a judge's robes and wig, and a mortarboard and gown – all signs of authority figures that perhaps recur in Jonzi D's dreams. They and the British flag, the ultimate sign of the nation state, surround and cover/smother the body underneath. They situate and ground Jonzi D's black British identity.

In the 1997 video record of *Aeroplane Man* Jonzi D begins his rap enacting a racist incident, playing the racist himself. Standing with his hands on his hips and a puffed up chest, he shouts aggressively in a strong cockney accent 'Oi, come over 'ere, take all our jobs, all our 'ouses, all our women. Fuck off back to your own country or I'll serve you up mate', ending with a terrifying Nazi salute which remains frozen in his body for several seconds. The performance graphically shows one of the effects of Jonzi D's black British identity. In this sense *Aeroplane Man* is a performance that, as Gilroy argues, 'can be used to create a model whereby identity can be understood neither as a fixed essence nor as a vague and utterly contingent construction to be reinvented by ... will and whim ... Black identity ... is lived as a coherent (if not always stable) experiential sense of self' (1993: 102). The performance of the racist incident and the signs of authority which surround and cover Jonzi D demonstrate that black identity is not a 'contingent construction' but that it is lived as an 'experiential sense of self'. In *Aeroplane Man*, between enacting the characters encountered on his travels, Jonzi D comments on them as himself. Going back and forth between the characters and himself he shows that his British identity is partial, fragmented and fluid 'but not without some sort of grounding in individual sociohistorical circumstances' (Gilbert, 1995: 344–5). When Jonzi D performs as himself the audience is reminded of the sociohistorical circumstances which prompted his journey in the first place. His encounters with people and cultures exhibiting remnants of different histories on his travels also show that identity making processes can be painful, leaving scars, because these identities have 'been enforced by the enduring memories of coerced crossing experiences like slavery and migration' (Gilroy, 1996: 20).

Identity, place and displacement

The roles of place and displacement in the construction of identities which have been shown as central in *Aeroplane Man*, are also evident in *Ellis Island*. The spaces of Ellis Island are foregrounded in the film indicating their importance. The work's title identifies its subject matter as the immigration centre which is part of American history and has a 'special place in the national psyche', as an article in *World Architecture*, describing its recent restoration as a museum, stated (Vickers, 1991: 62). The title sequence of the film shows Ellis Island, off Manhattan, ironically not far from the Statue of Liberty. As the *World Architecture* article states: 'if the design of the Statue of Liberty was intended to set the spirits soaring, then the architecture of Ellis Island was meant to project a more circumspect blend of civic welcome, federal gravitas and simple logistical practicality' (ibid: 63). Throughout the film several slow panning shots of the decaying buildings hint at the former grandeur of the centre and its role. Its scale is evident in shots of high walls and windows, vaulted ceilings, large barn-like spaces, pillars and long corridors. A commentary informs a contemporary tour group that there were 33 buildings on the island covering 599 575 square feet. This was clearly a vast enterprise. The imposing size of the centre was a sign of its institutional authority and an indication of the extent of its function; the commentary states that at its peak it processed over 11 000 immigrants in one day.

The film shows the abandoned spaces inside the building with piles of desks, filing cabinets, a wheelchair and some crutches. Together with nineteenth-century white tiling these objects construct the space as a particular institutional place – a clinic or sanatorium. The film adds to this when the contemporary visitors are seen at one point wearing surgeons' masks, as if to protect them from contamination. The centre's position on an island constitutes it as a place of isolation and containment from which it is difficult to escape. Foucault identifies the role of isolation in systems of power, indicating its facilitation of intensification and consolidation.

Monk and Panzen's filming in the Ellis Island detention centre, including shots of contemporary visitors touring the site, underlines the importance of the site itself as a remnant of its past that still resonates with the present. The ways in which the film lingers on details such as the architecture, floors and walls continually emphasize the significance of these places in the construction of the identities of those who passed through them. This treatment of space no doubt arises from Monk's claim that when she visited the island she 'sensed the spirits of people still in those rooms' (in *Ellis Island* publicity, 1981).

Technologies of power – subjectification, normalization and examination

By exploring the treatment of identity, difference and power in *Aeroplane Man* and *Ellis Island* some of the ways in which people are subjected have come to light. Foucault, whose objective 'has been to create a history of the different modes by which . . . human beings are made subjects' (1982: 208), has developed a schema of three modes of objectification of the subject: 'dividing practices', 'scientific classification' and 'subjectification' or specifically 'the way a human being turns him, or herself into a subject' (ibid). All three are apparent in *Ellis Island*. The film's construction and deconstruction of these modes of subjectification is explored in the light of Foucault's theories of disciplinary technologies of power, such as normalization and examination.

Foucault's theories derive from his detailed analyses of the operations of power in particular historical contexts and periods. They explain how subjects are constructed and construct themselves while revealing how the mechanisms of power at work in these processes operate. 'Subjectification' paradoxically 'denotes both the becoming of the subject and the process of subjection' (Butler, 1997: 83). The becoming of the subject through processes of iteration and repetition allows for resistance. 'Foucault formulates resistance as an effect of the very power that it is set to oppose' (ibid: 98). In other words the disciplinary apparatus that produces subjects 'brings . . . the conditions for subverting that apparatus' (ibid: 100). *Ellis Island* demonstrates the relevance of these theories in a contemporary context. Seen in conjunction with a work like *Aeroplane Man* it illustrates that Foucault's theories can reveal the workings of power that construct 'racial' and national differences and the ways in which they pervade and are perpetuated through networks of legislation and governance, but also allow for resistance. Much of Foucault's work concerns the workings of power on individual bodies, which is why his theories, which have been frequently applied to other areas of cultural practice, are particularly pertinent for dance.

Subjectification

'There are two meanings of the word *subject*,' Foucault writes, 'subject to someone else by control and dependence, and tied to his own identity by a conscience or self-knowledge. Both meanings suggest a form of power which subjugates and makes subject to' (1982: 212). The opening of *Aeroplane Man* works by allusion to suggest Jonzi D's status as a British subject and how this status relates for him to figures and institutions of authority and control such as the state, the law and education. It suggests ways in which he is subject to someone else by control and dependence illustrating Foucault's first meaning of subjectification. Throughout most of the rest of *Aeroplane Man* spectators witnesses Jonzi D exploring his identity and subjectivity which are constructed through travels prompted by his conscience or self-knowledge, illustrating Foucault's second meaning of subjectification. In *Ellis Island* Foucault's first mode of subjectification is only too apparent as the immigrants are subjected to various technologies of power such as measurement, classification and medical examination by the authorities.

When they are named and registered they become legal subjects. There is also evidence of the immigrants colluding in some of these processes, for example when one of the immigrants measures himself. This is an instance of Foucault's second meaning of subjectification involving self-discipline and self-subjectification.

Foucault terms the workings of power 'disciplinary technologies' and as the American postmodern geographer Edward Soja indicates, these 'operate through the social control of space, time and otherness to produce a certain kind of "normalization"'(1996: 161). The aim of disciplinary technologies, according to Foucault, is to forge a 'docile body that may be subjected, used, transformed and improved' (1977: 198), and space plays a role in this because 'discipline proceeds from an organization of individuals in space, and it requires a specific enclosure of space' (Rabinow, 1984: 17). The immigrants in *Ellis Island* are seen as enclosed and contained in that space facilitating their organization and subjectification. One of the roles space plays in the subjectification process in *Ellis Island* is evident in the filming and performance which evokes Foucault's notion of panopticism. Foucault cites the British philosopher Jeremy Bentham's (1748–1832) Panopticon; a circular viewing tower within prisons to ensure permanent surveillance of the prisoners whose cells surrounded it. Surveillance of this kind is suggested in *Ellis Island* when the camera slowly pans round a space in the abandoned building. Later a series of pans of another decaying room occur with a group of about 12 performers in contemporary black tops and trousers and dark glasses, repeatedly running across the space as a group in and out of the camera's field of vision. Some immigrants are also in the space but they appear unaware of the running group, who look as if they are trying to stay out of the range of the camera but fail to do so. The camera behaves like a preset surveillance camera as it pans back and forth stopping at regular intervals. The group dressed in black both emphasize and resist the surveillance by their bizarre performance, showing how the immigrants also resist the camera's normalizing gaze, as it is impossible to see what either group are doing when they are out of shot. In terms of the subjection processes at work, both groups evade them, by either repeatedly dodging in and out of view of the camera or by often being out of range. In this sense they are performing a 'repetition which does not consolidate . . . the subject, but which proliferates effects which undermine the force of normalization' (Butler, 1997: 93).

Various other spatial organizations of immigrants are presented in the enclosed space of Ellis Island. Some of these 'organizations' seem authentic, such as when immigrants are seen waiting at a quayside. Others are clearly constructed. For example, two parallel lines of five men, all dressed in black, some with hats on with arms crossed on their chests, placed centrally in the shot, are seen lying side by side on the ground. Other organizations fall in between these two categories, for example, when individuals are seen sitting on benches. By presenting organizations of people in space as a continuum in this way the film blurs the boundaries between what appears to be 'natural' and what is clearly not. This suggests on the one hand that these spatial organizations are constructed, providing evidence of imperceptible instruments of power at work, but on the other, that the blurred boundaries resist uniform subjectification, and instead allow for an appreciation of the positive values of difference.

Normalization

Normalization is a term Foucault uses to describe a process of measuring up to standards, of conforming to recognized codes or norms. It includes 'all those modes of acculturation which work by setting up standards or "norms" against which individuals continually measure, judge, "discipline" and "correct" their behaviour and presentation of self' (Bordo, 1993: 199). Measuring is presented as a theme throughout *Ellis Island* by the inclusion of archaeological black and white striped measuring sticks of different sizes (Plate 7). These first appeared in *Recent Ruins*, where 'generations of archeologists are presented . . . all . . . compulsively measuring' (West, 1980: 50). In *Ellis Island* initially the measuring rods are seen placed alongside or underneath objects, such as a wheelchair and a kidney dish, as if they are there to indicate size. Next they are seen in still shots of people posing for photographs. In each a stick is held against them by an arm in a white sleeved coat, as if measuring. One young man has a large stick thrown to him, which he catches and holds at his side looking up at it, measuring himself against it. These sticks operate as visual indicators of assessment. Foucault claims: 'the success of disciplinary power derives . . . from the use of simple instruments: hierarchical observation, normalizing judgement' (1977: 170). The immigrants are being measured or measuring themselves against the sticks, providing evidence of both of Foucault's kinds of subjectification. Juxtaposing images of people and sticks with images of objects and sticks suggests comparison. The measurement of the people objectifies them. This and the variable sizes of the sticks – some are a few inches long, others several feet long – deconstructs and exposes the operations of power at work.

This measuring theme is also rehearsed when shots of peoples' faces have measurements put on them; a line is drawn along a woman's forehead, above it '125 mm' is written. Parallels with archaeology are evident, where shards of bone or skulls are measured to determine their age. In the light of Foucault's theories, it seems that more can be read into this. The ways in which dimensions of the woman's forehead are being calculated and noted are suggestive of some sort of ethnic or 'racial' assessment and classification based on theories of evolution or eugenics. Foucault claimed that 'it was on the basis of the . . . rationality of social Darwinism that racism was formulated' (in Rabinow, 1984: 249). The woman is individualized and singled out as different from the others by this treatment, in the same way as the circle drawn round the nose of another woman and the letter J written next to it single her out as Jewish. The postcolonialist Homi Bhabha, discussing normalization in the context of colonialism, proposes, 'the natives are . . . "individualised" through the racist testimony of "science"' (1990: 76). The measuring, classifying and labelling depicted so graphically in *Ellis Island* are vivid portrayals of 'racial' individualization. Foucault argues, 'Discipline . . . "trains" . . . bodies . . . into . . . individual elements . . . separate cells . . . genetic identities. . . . Discipline "makes" individuals . . . as objects and instruments of its exercise' (1977: 170). The measuring incidents in *Ellis Island* are examples of Foucault's disciplinary technologies of power at work. The arms clothed in white coats that hold up measuring sticks to the immigrants and the hands that inscribe numbers and letters across people's faces belong to 'orthopaedists of individuality' in Foucault's terms (ibid: 294). They are part of a normalization process which involves an 'appeal to statistical measures and judgements about what is normal and what is not in a given population' (Rabinow, 1984: 21). A long list of statistics about Ellis Island – the results of measuring – is included later in the film in the contemporary commentary accompanying a guided tour.

Examination

Examinations of various kinds in *Ellis Island* provide further evidence of Foucault's subjectification and normalization processes at work. According to Foucault, 'the success of disciplinary power derives . . . from . . . a procedure that is specific to it – the examination' (1977: 170). The section on 'The Examination' in Foucault's *Discipline and Punish* begins,

> the examination combines the techniques of an observing hierarchy and those of a normalizing judgement. It is a normalizing gaze, a sur-

veillance that makes it possible to qualify, to classify, and to punish. It establishes over individuals a visibility through which one differentiates them and judges them ... the examination is highly ritualized. In it are combined the ceremony of power and the form of experiment ... it manifests the subjection of those who are perceived as objects and the objectification of those who are subjected.

(ibid: 184–5)

A scene in *Ellis Island* depicts a medical examination. In a room divided by curtained screens and containing rows of washstands, trolleys, a wheelchair and tables with bowls and clinical instruments, a man stripped to the waist and a woman clothed, stand and wait. Two white coated men with bowler hats, presumably doctors, enter together, and walk to face the waiting man and woman; their costume and unison entrance ritualizes this act, as does the entrance of a man and woman in contemporary black dress, who go and stand beside the waiting immigrants. They seem to act as witnesses and reflections or shadows as they watch and imitate the moves of their immigrant partners who appear unaware of them. The immigrant man is turned round by the doctor and made to bend forward while his back is tapped and prodded. The immigrant woman's head is tipped back by the other doctor who examines her eyes pulling up the lids until a bulging eyeball almost comes out of its socket. A uniformed figure passes through, surveying this spectacle of subjection. The repetition of the immigrants' manipulated movements simultaneously by their contemporary partners undermines and resists the examination that is occurring. This mimicry is supplementary; it happens without the doctors' manipulation. It is a contemporary trace of what happened retained from the past. It can be compared to the running of the similarly dressed group of performers in the 'panopticism scene' described above, in the sense that it is subversive because it is excessive. Both incidents trouble any suggestion that *Ellis Island* is a straightforward historical reconstruction, they are part of Monk's deconstructive approach.

In another scene in *Ellis Island* a woman sits at a table trying to complete a puzzle, arranging black and white squares, a kind of 'intelligence test'. A uniformed official stands over her watching. After rearranging the squares and forming a pattern, the woman looks up to the official for approval. Here, as Foucault claims: 'the body is ... directly involved in a political field; power relations have an immediate hold upon it, they invest it, mark it, train it ... force it to carry out tasks, to perform ceremonies, to emit signs' (ibid: 25). The 'body' is judged according to how it performs, whether it measures up to certain standards. On the basis of her performance, the woman is accepted or rejected, she either passes or fails this examination. As Lingis suggests in an article discussing Foucault's theories, 'norms are produced by the comparison surveillance makes possible between the levels, abilities, and performances of different individuals' (in Welton, 1999: 292). The film leaves the result of this test open, there is no indication whether the woman is successful or not. She does not arrange the squares to form the most 'obvious' pattern, but another pattern. The immigrant has travelled from symbol systems of her own indigenous or peasant culture into the symbol systems of modern, scientific discourse whose 'logic' she cannot fathom. It seems that Monk is being playful here. The woman is being imaginative, exhibiting a form of intelligence that the test cannot measure. In this sense the woman is resisting normalization.

The operations of administration, documentation and organization depicted in *Ellis Island* are further examples of Foucault's disciplinary technologies of power; for example, the official who asks an immigrant's name, writes in and rubber stamps a ledger. Immigrants in the film are seen with numbers pinned to their clothing. Much of *Ellis Island* consists of 'snapshot images', tableaux of immigrants posing for photographs and shots of faces with measurements and labels like SERB and 'J' for Jew written across them. All of these administrative processes 'subjugate' the immigrants 'by turning them into objects of knowledge' (ibid: 28). They become 'cases' for documentation, statistics for dossiers. As Foucault argues: 'the examination, surrounded by all its documentary techniques, makes each individual a "case", a case which ... constitutes an object for a branch of knowledge and a hold for a branch of power' (ibid: 191).

By depicting examination and documentation processes *Ellis Island* exposes 'a policy of coercions that act upon the body ... a machinery of power that explores it, breaks it down, and rearranges it' (ibid: 138). Bodies are explored by 'doctors', broken down into parts when examined or labelled, and rearranged for numerous photographs in *Ellis Island*. These disciplines produce subjected and practised bodies, 'docile bodies' in Foucault's terms. The marks of these disciplinary practices of subjectification are still apparent as, in another context, *Aeroplane Man* indicates. Jonzi D's visible sign of difference from white English people – his skin colour – results in a whole gamut of associations, connotations and labels, which operate to make him feel displaced in the place where he was born. These have in part been perpetuated by and are the legacy of administrative practices of detention centres like Ellis Island.

The role of language in operations of power

In *Ellis Island* and *Aeroplane Man* language is inscribed with power and contributes to the displacement experience. The use of language to label in *Ellis Island,* and the translation and transformation that occurs in the recording of names, demonstrate language's 'power to name, identify, classify, domesticate and contain' which 'simultaneously doubles as the power to obliterate, silence and negate' (Chambers in Chambers and Curti, 1996: 48–9). When only the back of an immigrant's head is seen as he gives his name, and when his name is changed radically, he is effectively being 'obliterated' and 'negated'. The lack of a 'natural soundtrack' accompanying the immigrants' scenes both situates the film temporally alongside silent movies, and 'silences' the immigrants – they are given no voice. This is underlined by the contrast with the 'contemporary scenes', which have 'normal' sound, and with a language lesson, where only the teacher speaks.

This language lesson in *Ellis Island* shows that the '"official" normative language of colonial administration' is also the language of 'instruction' (Bhabha, 1990: 73). The immigrants mouth after a teacher words such as 'vacuum cleaner', 'Empire State Building', and 'microwave'. The teacher's words are heard on the soundtrack but the immigrants' are not, they are silenced. Discussing the role of language in colonialism, Tarasti suggests, 'very often the subordinated voices can speak – have their voices heard – only after they have adopted the langue of the dominant culture' (1999: 75). This language lesson is another example of Foucault's disciplinary technologies of power deployed to subjectify the immigrants. This class of immigrants sitting behind desks obediently repeating words, is an example of Foucault's 'docile bodies' being 'transformed' and 'improved'. By including the Empire State Building, not just a national monument, but at one time the tallest building in the world, the ultimate sign of American phallic power, *Ellis Island* suggests that this is not only a language lesson. It is also a presentation of the dominant power's culture and achievement, designed to impress, subdue and, in the process, oppress. By including words such as 'microwave', which did not exist when Ellis Island was operative, the film is time travelling, or flattening history, showing connections between past and present, and playfully poking fun at what is going on, deconstructing the constructed nature of this performance.

The power of language to displace is also evident in *Aeroplane Man* when Jonzi D is told in no uncertain terms to 'Fuck off back to your own country' and to 'fuck off outa here'. The language and form of the performance poetry of rap provide a platform for the uncompromising honest expression of, as Jonzi D himself puts it, 'how I see the world' (1997). Rap is part of a long tradition of black performance which has often been overtly political. As Gilroy points out: 'the interface between black cultural practice and black political aspirations has been a curious and wonderfully durable ... phenomenon' (1995: 12). The form of rap which cleverly plays with words and rhymes in a very vivid manner is 'all about the power and use of the word' (Gottschild, 1996: 134). Whereas dominant language subjectifies, as is evident in *Ellis Island,* the poetry of rap undermines and subverts the dominant language, as does cockney rhyming slang, which Jonzi D also incorporates in his performance. As the Russian literary theorist Mikhail Bakhtin, whose ideas are discussed in relation to dance in Chapter 9, has indicated, all language is ambivalent but officialdom often tries to close it down. Rap and other forms of resistance seek to restore the ambivalences and keep language alive and open.

The rhythm of rap is also a key element in *Aeroplane Man*. Rap combines 'rhythm and text with the ideology of power' (ibid: 137). The force of the rhythm combined with the uncompromising use of language can be very powerful. Rhythm drives home the point of the poetry, it mobilizes an audience. Rhythm is 'a component [in rap] that can inspire fear in a Europeanist culture that knew enough about the power of African rhythm to prohibit drumming by enslaved Africans' (ibid). That power is evident when Jonzi D performs. I witnessed a predominantly young, black audience cheering, yelping, hollering and joining in when he performed. As one reviewer has stated, 'boundaries between performers and spectators are constantly dissolved in Jonzi's work as he speaks directly to them, expecting responses (which he gets)'(Leask, 1998: 47).

Another infectious characteristic of rap language and performance, which *Aeroplane Man* demonstrates, is the use of comedy. Jonzi D's impersonation of characters from the Nazi racist to laid back Caribbean and American streetwise dudes is cleverly and carefully observed and humorously performed. He is a talented mimic whose portrayals of character are often hilarious. There is a comic moment in the extended version of *Aeroplane Man* when he tries to teach the 'foreign language' of cockney rhyming slang to a couple of 'New York chicks'. This incident humorously demonstrates the subversive force of mimicry while also celebrating difference. Jonzi D's performance is evidence that 'hip-hoppers... are both "gangstas" and clowns' and that 'the hip-hopper is the latterday incarnation of the trickster, that dangerous inscrutable enigmatic quotient in African religions' (Gottschild, 1996: 137, 138). If 'the question for the post-colonial artist is how to speak in a language which belongs to the colonizers and yet represent the viewpoint of the colonized' (Betterton, 1996: 168), Jonzi D's answer, evident in *Aeroplane Man*, is to use the performance poetry of rap.

Spatial containment, borders and territorialization

Important parallels can be identified between the containment and control of the immigrants on Ellis Island and Jonzi D's experience of displacement in Britain in the nineties. Referring to Britain, Bhabha describes the 'entertainment and encouragement of cultural diversity' as a form of control and 'containment'. He argues, 'a transparent norm is constituted... by the host society or dominant culture, which says ... "these cultures are fine, but we must be able to locate them within our own grid". This is... a *creation* of cultural diversity and a *containment* of cultural difference'[his emphasis] (1991: 208). *Ellis Island* shows that the immigrants' containment involved marking them as different and othering them. The processes of measurement, classification and examination they underwent could be seen to be locating them within a 'grid'. The opening moments of the extended version of *Aeroplane Man* show Jonzi D also placed or located within a 'grid'. His cultural difference renders him in a particular relation to the nation state and its institutions of authority, represented in the performance as a form of containment by the duvet covering him and the three piles of clothes placed in a triangle around him. Spatial containment or segregation is shown to be fundamental to experiences of displacement in both works. *Ellis Island* and *Aeroplane Man* show that 'segregation reproduces itself: spaces of otherness become not only repositories of "others" but... one of the primary indicators/producers of alterity' (Schick, 1999: 44). The opening scene of *Aeroplane Man* suggests that the racist practices that are portrayed as part of the administrative infrastructure in *Ellis Island* have become further institutionalized within the nineties British state apparatus. As Soja argues: 'hegemonic power... produces difference as a . . . strategy to create and maintain modes of social and spatial division that are advantageous to its continued empowerment and authority' (1996: 87).

Spatial division and segregation are created and maintained by borders. *Ellis Island* reveals some of the investment put into the maintenance of borders by an immigration control centre. Yet borders are social constructions. They are 'arbitrary dividing lines that are simultaneously social, cultural and psychic', they create 'territories to be patrolled against those whom they construct as outsiders, aliens, the Others . . . places where claims to ownership – claims to "mine", "yours" and "theirs" – are staked out, contested, defended and fought over' (Brah, 1996: 198). In Jonzi D's rap, territorial issues of what is "mine" and "yours" are raised at the outset by the British Nazi racist who accuses Jonzi D of 'coming over 'ere, taking all our 'ouses, all our jobs and all our women'.

Territorial issues are further raised on Jonzi D's travels. Each time he crosses a border into another country he is perceived as an outsider and told to return to England. For Jonzi D, as *Aeroplane Man* shows, both in Britain and elsewhere, borders create what the postcolonialist Edward Said has termed 'the perilous territory of not-belonging' (1990: 359). In the *Aeroplane Man* video interview Jonzi D talks about displacement as 'that feeling of not being welcome, of not being comfortable of not belonging' (1997). Said consistently argues for 'a world in which traditional boundaries of all kinds are to be questioned' since they are 'often nonsensical' and 'oppressive' (Kasbarian, 1996: 531).

While demonstrating different aspects of oppression associated with national borders, *Aeroplane Man* and *Ellis Island* also challenge and question boundaries in empowering ways. Neither Jonzi D nor the *Ellis Island* immigrants 'fit' easily into a nation state. By not 'fitting' they show up the rigidity of the institutional and cultural apparatuses that serve and are perpetuated by the contained and bounded nation state. Jonzi D and the immigrants can be seen to be both 'inside' and 'outside' at the same time. Jonzi D is both inside and outside England – he was born and bred in England but he is not accepted. He is also both inside and outside the places he visits – inside because he is black, he owns, embraces and can perform aspects of their culture, and outside because he is English. The immigrants are inside the American immigration centre but they are treated as outsiders. They don't 'fit' when they fail to complete the intelligence test in the 'normal' way, their names don't fit, but they also exceed the conceptual boundaries when their contemporary black-clad 'shadows' repeat their movements and evade the surveillance camera's normalizing gaze. When a group of immigrants are seen waltzing with each other in their traditional nineteenth century peasant dress, the dance doesn't fit the costume – the sign is split. When a group of immigrants poses for a photograph, for a moment they are fixed and frozen, they look like a photograph, then someone blinks and a woman brushes a speck of dust off a coat lapel. They have exceeded conceptually the boundaries of the camera shot. All these excesses or supplements are instances of resistance that both challenge the normalizing gaze and the constructed boundaries. They show that the borders or boundaries *require* these 'others' in order for them to exist. The outsiders define the borders of what is inside. They point up the arbitrary and constructed nature of borders and how they can be challenged.

Conclusion

This chapter has shown how *Ellis Island* and *Aeroplane Man* examine issues of identity, difference and power, and demonstrate their complexity. The works reveal that identity is not a monolithic, static concept, with identifiable, fixed origins, but that it is a process or mode of differentiation that is changeable and bound up with systems of power that operate spatially. Both *Ellis Island* and *Aeroplane Man* demonstrate how particular in-between spaces and places play a part in the construction of identity. The journeys of Jonzi D and the immigrants through these liminal spaces show them crossing literal and conceptual thresholds in their rites of passage. Existing both inside and outside these in-between spaces, and at times exceeding them, effectively blurs boundaries and challenges notions of a rigid, self contained subject with a fixed identity. The blurring of boundaries is also evident in the way both pieces, by employing several different media, demonstrate what Monk terms 'a mosaic way of perceiving' (in Zurbrugg, 1993: 98). In *Aeroplane Man* the in-between, unfixed nature of the improvised performance format, resulting in no one performance being the same, also troubles notions of fixity. Both *Ellis Island* and *Aeroplane Man* blur boundaries to suggest fluid subjectivities and identities that celebrate differences.

In both works 'themes of identity have been explored ... through the relation of personal to historical memory, through journeys, both real and metaphorical, and through the representation of self from the point of view of those displaced from the "centre" by ... race' (Betterton,

1996: 193). Jonzi D's graphic portrayal of his experiences of displacement demonstrates that the effects of the technologies of power that subjected and individualized the immigrants as different in *Ellis Island* are still felt today. Both works through their portrayal of displacement arising from multiple historical and contemporary crossings of the (black) Atlantic show how 'here' informs 'there', how 'we' inform 'they' and how 'then' informs 'now'.

Part III
Inside/Outside Bodies and Spaces

8
Fleshy Corporealities in Trisha Brown's *If You Couldn't See Me*, Lea Anderson's *Joan* and Yolande Snaith's *Blind Faith*

Introduction

While the focus in the last three chapters has been on actual and metaphorical indeterminate hybrid in-between spaces, the concern in this and the next chapter shifts to bodies, specifically the actual and conceptual boundaries of bodies where bodies meet space, and where inside and outside are difficult to distinguish.

This chapter focuses on the choreography of the inside/outside borderlines where bodily flesh, fluids and folds meet space in *If You Couldn't See Me* (Brown, 1994), the dance film, *Joan* (Anderson, 1994) and *Blind Faith* (Snaith, 1998). The American choreographer Trisha Brown's solo choreography for her naked back in *If You Couldn't See Me* directs the audience's attention to the flesh, folds, muscles and bone structure of this relatively unfamiliar body part (Plate 8). *Joan*, also a solo for the British choreographer, Lea Anderson, is inspired by Carl Theodor Dreyer's (1928) film, *The Passion of Joan of Arc*. Both works focus on Joan's spirituality merging with her fleshy corporeality. The British choreographer Yolande Snaith's *Blind Faith* is inspired by the work of Leonardo da Vinci, particularly *The Last Supper* (1498) and by Renaissance anatomical experiments evident in paintings such as Rembrandt's *The Anatomy Lesson of Dr Tulp* (1632) (Plate 9).

The materiality of corporeality is central in these dances. In *If You Couldn't See Me* spectators' attention is drawn to the moving flesh, musculature and skeletal structure of Brown's back by her low backed costume, by side-lighting, the plain darkness that surrounds her, and the fluidity of her loose-limbed movement style. In *Joan* the materiality of corporeality is foregrounded through close-ups of 'Joan's' face and head, the use of the camera to get 'inside' her body, and through video special effects. Bodily flesh, fluids and folds are emphasized in *Blind Faith* through choreography based on contact improvisation and through the dancers' portrayal of the investigation through manipulation of the near naked matter of bodies. In all three works there is much imagery of folding, and in *Joan* and *Blind Faith*, of fluid and fleshy matter such as water, wine, blood, tears, saliva and bodily innards. All pieces also make extensive use of light to highlight fleshy surfaces and to evoke mystical, sometimes trance-like moments or transformatory bodily experiences.

If You Couldn't See Me is a movement based piece with no obvious references to anything outside of itself. The focus is Brown's choreographic exploration of the performance potential of her back. The visual theatre style of *Joan* and *Blind Faith* is very different. There are references to painting, sculpture and film – both Anderson and Snaith had a visual art training[1] – and the pieces are historicist and include religious references to body/soul relations and the mortality of the body. However, whether through imagery that is movement based or in a visual theatre style, all three dances explore anti-dualistic ideas that focus on female subjectivity.

The dualism of the French philosopher René Descartes (1596–1650), sees the mind and body as separate entities where the body materially occupies space and is a container for the conscious mind. From this philosophical perspective, outlined in the Introduction and examined in Chapter 10, perception is organized around a series of binary oppositions such as mind/body and self/world, where the first of the pair is associated with the masculine and valued over the second which is associated with the feminine. By putting the body at the heart of their explorations, the dances are reinstating it and its associations with the feminine as central to subjectivity. In the dances central female figures in different ways infuse the works with particular kinds of embodied energy. I argue that there are resonances between this energy and that theorized by the French post-structuralist Gilles Deleuze as inherent in the fold, and that the focuses on flesh I identify in the works can be informed by the French based Bulgarian feminist Julia Kristeva's theories of abjection associated with the feminine. These theories are explored, together with the American theorist Susan Bordo's feminist account of Cartesian philosophy, to aid analysis of the dances. They all allow a focus on the body and subjectivity from an anti-Cartesian perspective, although the epistemologies of Deleuze and Kristeva have very different bases. Deleuze critiques psychoanalysis whereas Kristeva draws explicitly on it, but they are both interested in new, non-fixed identities open to otherness, making their work pertinent to this analysis.

After an introduction to the theories of Deleuze, Bordo and Kristeva and to each of the dances, the chapter focuses on the 'flesh', 'fluids' and 'folds' in the dances. The theories are explored in these contexts to point to ways in which the dances can suggest possibilities of rethought embodied subjectivities associated with the feminine.

The theories of Deleuze, Bordo and Kristeva

Gilles Deleuze's theories of 'the fold' are derived from his radical conception of the Baroque read through an interpretation of the seventeenth and eighteenth-century philosopher Gottfried Wilhelm Leibniz. Deleuze draws on Leibniz because he 'was the first thinker to "free" the fold, by taking it to infinity' and on the Baroque because it 'was the first period in which folding went on infinitely' (Deleuze, 1995: 159). The Baroque style in art, which straddled the seventeenth and eighteenth centuries – evident in painting, sculpture and architecture particularly in Italy – was developed to act on the emotions of the spectator, conveying, for example, the agonies and ecstasies of the saints. It did this through creating an illusion of movement through light and colour effects on folds of cloth, figures and flesh, to express profound and passionately felt religious emotions. Deleuze claims, 'without the Baroque and without Leibniz, folds wouldn't have developed the autonomy that subsequently allowed them to create so many new paths' (ibid). Key characteristics of folds that Deleuze explores and that suggest 'new paths' are the energy and force inherent in folding that 'spills over' infinitely resulting in movement bound up in the form of folds.

Deleuze's theories of the fold work on many levels. I am exploring Deleuzian folds which are conceptual alongside dances where many of the folds are actual and evident in the dancing bodies. Deleuze, however, derives his ideas from matter citing the body as one possible source (1993: 34). He explores Leibniz's ideas of perception in part through the body and its relation to the world emphasizing sensuality over sole dependence on the visual. This contributes to the critique of Cartesian body/mind separation which is fundamental to Deleuze's position. The sensual characteristics of folds derived from the Baroque, which challenge the separations inherent in Cartesian dualism, focus on multiplicity, excess, connectivity, particularly that of bodies and souls, and a wave-like force, energy or movement. Productivity and multiplicity of folds are evident when Deleuze writes of 'a proliferation of principles' where 'play is executed through excess and not lack' (ibid: 67–8). These all resonate with the foregrounding of flesh, fluids and folds in the dances as do the fluidity of matter and the elasticity of bodies which are fundamental notions of Leibniz's philosophy. Another key characteristic of the Baroque explored in *The Fold*, which is also evident in the dances, is mystical experience. Deleuze writes of folds conveying 'the intensity of a spiritual force exerted on the body' (ibid: 122). In *Joan* and *Blind Faith* trance-like, ecstatic bodily states occur which are also excessive and seem to be associated with the mystical, 'what Deleuze . . . might call an event . . . the virtual sensation of a somatic moment of totalization and dispersion' (Conley in Deleuze, 1993: xii). The emphasis on the somatic is key. It is characteristically corporeal, a bodily experience, a 'mystical adventure' that 'convinces because no language can be said to represent what it means' (ibid: xii). This is why it is pertinent to explore the theory alongside dance.

Susan Bordo, like Deleuze, explores anti-Cartesianism. Parallels with his position are evident in her citing of Leibniz, her attention to multiplicity, and her recognition of the importance of the unity of subject and object in the pre-Cartesian medieval world, which can be likened to Deleuze's ideas of body/soul connectivity. However Bordo emphasizes the *gendered* nature of these ideas. She shows how historically Descartes' philosophy resulted in a 'masculinization of thought' involving a 'flight from the feminine' (1987: 9). This meant a 'separation from the organic female universe' (ibid: 5) through a 'transcendence of the body' (ibid: 8). Bordo shows that the body's role within knowledge in medieval and Renaissance philosophy was overridden by seventeenth-century scientific and intellectual revolutions. The connections she makes between a Cartesian masculinization of thought and transcendence of the body aid an understanding of the ways in which a reinstatement of corporeality in the dances challenges Cartesian ideas and points to notions of subjectivity associated with the feminine. The loss of the medieval and Renaissance sense of 'being one with the world' (ibid: 106) and of any connectivity between subject and object is replaced by Descartes' philosophy of dualisms such as soul/body, self/other and subject/object. These binary oppositions privilege the concept of a rational, self-contained, unified subject that is considered ideal and associated with a traditionally white, male norm. Bordo claims that in Cartesian philosophy associations of the world with the feminine were characterized as evil and destructive and in need of suppression and control in order to ensure the objectivity of science. This desire to control matter as the object of science is evident in the genre of seventeenth-century anatomy lesson paintings referred to in *Blind Faith*.

These ideas of control over matter also resonate with Julia Kristeva's theories of abjection which concern the expulsion of those material aspects of corporeality considered to be unclean and improper such as food, bodily fluids and waste. Kristeva claims that the binary coding of the human subject in terms of subject/object, inside/outside and self/other depends on expulsion of this abject, unclean and improper matter. In other words if the body is seen as a container, as it is in Cartesian theory, messy substances which threaten to overflow the body and break its boundaries disrupt the order and control the container signifies. This sense of order and control is fundamental to the Cartesian notion of a rational, unified subject associated with the masculine, and it is dependent on the expulsion of the abject. The revelation of the abject in the flesh and fluids of *If You Couldn't See Me, Joan* and *Blind Faith* where it is foregrounded rather than hidden, points to the impossibility of complete expulsion and therefore of complete control and the consequent ambiguity of the construction of the subject in Kristevan terms. For as Kristeva argues, 'abjection is . . . ambiguity. Because, while releasing a hold, it does not radically cut off the subject from what threatens it' (1982: 9). Kristeva places the abject on the side of the feminine (ibid: 71), which is why her theories prove useful when analyzing these dances that are exploring notions of female subjectivity. Kristeva also posits connections between religious discourse, the arts and the abject that seem particularly pertinent to analyses of *Joan* and *Blind Faith*, because of their subject matter.

Deleuze's ideas of folds foreground connectivities which implicitly challenge Cartesian dualisms. Bordo identifies connectivities between body and soul, subject and object and self and world as part of the medieval and Renaissance views of the world rooted in the body and associated with the feminine. A return to seeing the world in terms of connectivities in this way points to a new kind of embodied subjectivity associated with the feminine. This notion of embodied subjectivity can be likened to Kristeva's ambiguous subject in process bound up with her theories of the abject. In these theories she identifies the roles that material aspects of corporeality such as bodily fluids and waste, often associated with the feminine, play when they excessively overflow and disrupt the order and control of the traditional Cartesian subject associated with the masculine. There are connections here with the overflow and excess evident in Deleuze's infinite foldings. So although the theories of Deleuze, Bordo and Kristeva are distinct, each bringing a slightly different perspective to bear on *If You Couldn't See Me, Joan* and *Blind Faith*, the ways in which they overlap and inform each other, when explored alongside the dances, challenge traditional notions of subjectivity and suggest new possibilities.

The dances

If You Couldn't See Me is a ten-minute solo choreographed and performed by Trisha Brown to music composed by Robert Rauschenberg, who also designed her costume; a white, low-backed dress (Plate 8). Central to the piece is Brown's back which is turned to the audience throughout. Although the performance includes throw-away swings and lunges of the limbs, and curves, tilts and suspensions of the torso in Brown's typical loose-limbed style, attention is focused on the moving surface of her bare back. It is side-lit, such that the differently angled fleshy surfaces of her musculature and bone structure stand out against the backdrop of plain dark curtains. At times her back seems otherworldly taking on a life all of its own; a host of different images are evoked. This focus on the back challenges the dominant single viewpoint that privileges the body's front in performance and everyday life. It disrupts expectations and the logic of visualization, which normally objectifies the female, dancing body (see Chapter 10). The body and its relationship with space are seen differently. Brown's choreography for herself, a woman approaching sixty when the piece was first performed, through its focus on this wall of flesh and muscle, suggests new ways of thinking embodied subjectivity while making statements about age, gender and identity.

Since making the solo, Brown has taught it to two male dancers, Bill T. Jones and Mikhail Baryshnikov, who have each performed it alongside her as a duet entitled *You can see us* (1995), the men facing the audience and Brown facing away.[2] This allows the audience to see on male bodies facing forward what would normally be hidden, while the female body remains turned away defying normal objectification. Brown had been reading about the use of women's bodies in the media when making the piece, and she was only too aware of the ambiguity of putting a woman's body on stage. By limiting the exposure of the female body in *If You Couldn't See Me* she felt she had taken ownership of the image (Feliciano, 1996: 11).

Joan, is a six-minute dance film made for Channel Four television, first broadcast on 19 August 1994, and directed by Margaret Williams. Lea Anderson introduces it as 'physically internal choreography'. It employs various imaging techniques to foreground and penetrate the body's flesh. Joan's body is entered by the eye of the camera/spectator, ripped open and set alight without any actual damage being done, alluding to its permeability and her mystical nature. Anderson, who performs alone as Joan, resembles the iconic image of this saint with her short boyish haircut and costume of chainmail armour under a black leather tunic. The myth of Joan and the peculiarly spiritual nature of her corporeality is explored. Joan is the centre of attention throughout, almost always in the middle of the frame and mostly filmed in close-up.

Joan opens with close-up shots of 'Joan's' ear immediately suggesting connections between her corporeality and her spirituality. Joan heard voices; her ear is her link to her God. Throughout, the film returns to close-ups of Joan's head, face and ear, the latter often framed by her hand, further emphasizing its importance. The film critic Pauline Kael suggests that in Dreyer's film enlargements of Joan resulting from close-ups are 'shockingly fleshly' (1970: 329). *Joan* is similarly 'shockingly fleshly'. 'Joan's' body is dissected by the camera as it enters her ear, revealing bloody, membranous inner caverns, and 'Joan' rips open her leather tunic exposing a digital image of her pulsating interior (Plate 10). These shocking revelations of 'Joan's' bloody flesh can be seen as examples of Kristeva's abject 'modes of corporeality', but in *Joan* the materiality of her body is foregrounded rather than expelled. Joan's relationship with her God/spirit is further emphasized through intense lighting effects; her head is brightly lit from above and behind, suggesting her mystical status. Imagery of corporeality and of spirituality alternate and overlap throughout the film. Joan is portrayed as being *in* her body but also intensely spiritual through closeness to her god.

Blind Faith is a 58-minute dance initially created for the stage for four men (Paul Clayden, Jovair Longo, Rick Nodine and Russell Trigg) and a woman (Snaith)[3] with a video record of one of its performances. Drawing on imagery from *The Last Supper*, a rectangular banqueting table, designed by Barnaby Stone, is a central prop around and on which the performance occurs. The dance opens with Trigg's limp, Christ-like body, wearing only black trunks, draped over the table like a corpse. The other dancers process around the table, ceremoniously place containers of water, a napkin and bread on the table, wash and dry Trigg's foot and place the bread on his body, which the three men then lengthily examine and manipulate. To Graeme Miller's accompaniment of repetitive chanting, Snaith performs a dynamic solo where she appears to be summoning up mysterious powers, she rolls on the floor and rises up onto her knees, her arms outstretched heavenwards. The 'corpse' 'comes to life' and dances its own macabre solo with postures reminiscent of the Renaissance anatomist Andreas Vesalius's famous drawings (Plate 11). Dances suggesting compulsive and perhaps transformational states are performed, where all the dancers walk around the table progressively increasing speed, imitate Snaith jumping on the spot repeatedly, and finally gaze as if transfixed into lights shining up through the table. Ecstatic and trance-like dances follow, one culminating in Clayden having a kind of fit, where he spits on the table. A series of tableaux of near naked bodies are created on the table (Plate 12) emphasizing the visual theatre style of the piece, also enhanced by special lighting effects including a panel of lights over the table as in an operating theatre, and the use of blue and gold washes which transform the appearance of the flesh of the near naked bodies from cold to warm, dead to live. These effects, when combined with tableau-like poses of groupings of bodies are reminiscent of imagery from Renaissance and Baroque, often religious, paintings. The dance ends with tableaux from *The Last Supper*, characterized by Baroque-like, flowing and flame-like postures and gestures of the 'disciples' culminating in a reconstruction of the painting with Snaith in Christ's position.

Blind Faith centres round investigations of corporeality. The whole dance seems to focus on bodies and their materiality. It does this in the context of particular historical, philosophical and religious ideas evident in strong visual tableau references to Renaissance and Baroque art. Corporeality is foregrounded through a focus on anatomy at the outset in visual references to anatomy lesson paintings, and reiterated through references to Vesalius's and Leonardo da Vinci's anatomical drawings which emphasize the flesh and musculature of the body. There is a sense in which the dance moves from 'flesh made word', evident in anatomical references, to 'word made flesh', apparent in *The Last Supper* tableaux.

Flesh

Flesh is immediately foregrounded in *If You Couldn't See Me* since Trisha Brown's naked back is at the centre of her performance. This relatively large expanse of flesh is rivetting precisely because audiences are not used to having this part of the body as the main focus of a dance. The fluctuation of the back's flesh, muscles and bones – 'the enigmatic language of the faceless body' (Sulcas, 1995b: 38) – becomes a dance on its own framed by movements of limbs and head. Steve Paxton, who has worked with Brown and known her over a thirty five year period since they performed together in the Judson Dance Theater (see Chapter 3), wrote at some length about *If You Couldn't See Me* in a letter to Brown published in *Contact Quarterly* (1995). Given his informed perspective, Paxton's letter is treated as a special source of particularly insightful information about the dance.

Paxton suggests that focusing on the flesh of Brown's back gives the body transformative potential, in the sense that a range of different images can be read into it. He describes Brown as 'a woman with a whole scene on her back' (1995: 94). He writes,

> the . . . revelation of your spinus erectae, scapulae and delicate pearls of spinal processes in the lumbar . . . sent a gasp through the audience. You cannot know . . . what the sculpture of your back can accomplish . . . it became an abstraction shifting from an anatomical event with muscles like a whippet to a large looming face to a mask – something alien and frightening.
>
> (ibid)

This description highlights the transformatory nature of the choreography and 'something alien and frightening' suggests mystical, otherworldly qualities. He continues, 'our eyes see . . . the moonscape of your back, the chiaroscuro, which as you raise your arms, fascinates us, bringing us closer and rendering you enormous' (ibid). The importance of light emphasizing the texture of the back's flesh and folds is apparent. Deleuze mentions this in relation to Baroque folds stating: 'the fold of matter or texture has to be related to . . . light, chiaroscuro, the way the fold catches illumination' (1993: 47).

Brown continually transforms, disassembles and reassembles herself throughout. Paxton recognizes this when he writes of 'the figure-ground flipping . . . as the dance progresses' (1995: 94) and when he concludes his statement about Brown's back being 'something alien and frightening' with 'although [it is] weirdly comic as the rest of the body reconnects and we see again a back' (ibid). The ways in which the back comes in and out of focus, appearing to change from one thing to another, underline its qualities of elasticity, flexibility and instability. Discussing the pliability of folded forms, Deleuze suggests, 'there's nothing more unsettling than the continual movement of something that seems fixed. In Leibniz's words: a dance of particles folding back on themselves' (1995: 157).

The transformations inherent in this fleshly choreography can effect a blurring of the inside/outside boundaries of the body. The muscles and bones of Brown's back, normally thought of as *inside* the body, are revealed. As one critic wrote, *If You Couldn't See Me* 'exposes Brown's articulated muscles ... movements ripple out into her limbs from deep inside her torso' (Felciano, 1996: 11). Connections are made between inside and outside, between deep inside the torso and the apparently external movements of limbs. This perception of the body and its movements challenges traditional notions of the body as a container of flesh and muscles, a controlled, Cartesian body, clearly articulating its borders, like that of a classical ballerina. Brown's fluid, loose, almost messy, movement style articulates connections rather than containment.

The particular relationship Brown's dancing has with space is foregrounded in *If You Couldn't See Me* because the dance is a solo, the first Brown has made in twenty years. Since there is no one else to dance with, it is as if 'space ... becomes this other body, accompanying ... Brown's dance' and 'Brown's choreography expands the notion of a choreographed body into the body of space' (Lepecki, 1997: 19). This animation of the body/space interface blurs the inside/outside boundaries of the body and of space allowing for the possibility of a new fluid, unstable, embodied subjectivity. Paxton also recognizes the special relationship Brown's dancing body has with space as she appears to transform and reconstruct the space around her while dancing. He claims that because of the focus on Brown's back, the upstage space is enlivened and 'the black curtain is not just an expanse of dark background to contrast with your lit figure and push you towards us.... Your orientation declares it a dark vista ... it acquires a dimensionless depth' (1995: 94). He continues, 'this illusory upstage spatial pliability – from positive dancing figure on black to negative dancing figure in front of potentized unguessable empty space is deeply satisfying, mythic and wholly theatrical' (ibid). The alternating positive/negative images of Brown's dancing body further underline its flexibility and its potential for exceeding its boundaries and refusing to remain contained or fixed. Paxton's use of the term 'mythic' to describe this experience links *If You Couldn't See Me* to *Joan* and *Blind Faith* in which mythic qualities are also apparent.

Joan opens with '*JOAN*' in white letters flashing onto the centre of a black screen in rapid succession in time to Drostan Madden's heartbeat-like sound score. Immediately corporeal mortality is signified in this flashing title image. Three close-up shots of 'Joan's' head and ear follow, each interspersed with a plain black screen, also timed with the heartbeat soundtrack, like the pulsing signals of life on an electrocardiograph. This appearance and disappearance of first the word '*JOAN*' (flesh made word) and then 'Joan's' head (word made flesh) also suggest the mystical ambiguity of Joan's mortal status. Now you see her, now you don't. The camera focuses on 'Joan's' ear again, getting closer and closer, as in Dreyer's film where 'giant close-ups' revealed 'startlingly individual contours, features and skin' (Kael, 1970: 329). The image changes to inside a long curving paper tunnel, followed by shots that resemble an ear's interior, with red and white cavernous spaces traversed by pulsating fleshy membranes. This internal journey through the ear's caverns has affinities with anatomical drawings of folded flesh such as those of Vesalius, and with Deleuze's description of matter, as offering 'an infinitely porous, spongy, or cavernous texture . . . caverns endlessly contained in other caverns . . . pierced with irregular passages, surrounded . . . by . . . fluid' (1993: 5). A shot of a road tunnel follows and, when light is seen at the end of it, a close-up of 'Joan's' eye, with tears rolling down her cheek, replaces it. 'Joan's' body has been invaded through her ear, her link with her creator. Views of the inside of her ear and head have been constructed by film. The road tunnel seems to suggest an internal journey to her soul, her spiritual quest. When the camera appears to enter her head, distinctions between body and soul become unclear. This invasion seems to illustrate Deleuze's claim that 'souls are everywhere in matter' (ibid: 11). This is also evident when 'Joan's' head is filmed alternately in positive tones (a dark image on a white ground) and in negative tones (a luminous white image transluscent like an X-ray against a golden ground, reminiscent of the mystical imagery of haloed saints). These plays with the image of her head, including an X-ray shot, blur inside/outside boundaries and the body/soul dualism suggesting the closeness of 'Joan' with her god. There are parallels here with Paxton's description of *If You Couldn't See Me* moving 'from positive dancing figure on black to negative dancing figure in front of . . . empty space' which he saw as 'mythic and wholly theatrical' (1995: 94).

When 'Joan' rips apart the front of her leather jerkin to reveal a computerized rendition of an abject fleshy pulsating interior (Plate 10), any expectations of objectification associated with disrobing are denied. The subject/object binary is disrupted, opening up possibilities for 'Joan' having a new identity, where inside and outside spaces of the body coexist in one image. This illustrates the ambiguity of the abject which is 'the space between subject and object' (Gross, 1990: 94) that is 'undecidably inside and outside the body' (ibid: 90). Deleuze identifies the connectivity of inside and outside spaces as a trait in the Baroque, where he argues: 'the infinite fold . . . moves between matter and soul . . . the outside and the inside' (1993: 35). The connotations of this hybrid fabrication of 'Joan', through deeply religious experiences of immortality, seem particularly apt given Joan of Arc's mystic status. Simultaneously, however, the image of 'Joan's' pulsating interior, exposed and therefore also exterior, can be read as demystifying 'St Joan', showing that she consists of flesh and blood just like other mortals. This illustrates abjection as 'the expression of both a division (between the subject and its body) and a merging (of self and Other)' (Gross, 1990: 92). The image encapsulates the fluidity of identity showing it is possible to be more than one thing simultaneously. This is also apparent in the transformatory choreography of *If You Couldn't See Me*. Deleuze suggests in *The Fold* that body and soul are both one and the same while also being different, that sameness and difference can occur simultaneously in one entity. He argues, 'everything is always the same thing... and... everything differs' (1993: 58).

This fluctuating coexistence of sameness and difference is also evident in *Blind Faith*. In its opening moments the flesh of Russell Trigg's body, laid out as if for dissection, suggesting resonances with Rembrandt's *The Anatomy Lesson of Dr Tulp* is foregrounded in various ways. The foot washing, performed over a bright white light shining up through the glass table top, focuses attention on Trigg's luminous flesh. The materiality of his corporeality is also alluded to when bread is placed on his torso and the water in the glass bowl magically turns red when Snaith drops powder into it. The juxtaposition of Trigg's Christ-like body with the bread and water turned into wine, has associations with the Eucharist of the last supper, an image to which *Blind Faith* returns at the end of the dance. When Trigg's body is lifted, folded and rolled over by the three male dancers and then 'dances', resembling in its postures and gestures Vesalius's anatomical drawings (Plate 11), it provides another reference to Rembrandt's multi-layered painting. Rembrandt had studied Vesalius's drawings and he possessed limbs anatomized by Vesalius (Sawday, 1995). The 'corpse' in *Blind Faith*, like that in Rembrandt's *Anatomy Lesson*, has become, in part, a mindless object of science. The flesh has become word, emphasized in the painting by the anatomy textbook open at the corpse's feet, at which some of the surgeons are gazing. This is a 'textualization' of the body (Barker, 1995: 72), a Cartesian body separated from its mind or soul, because of the 'conjunction between the Cartesian struggle of will and intellect, and Rembrandt's portrayal of the domination of intellect over the aberrant will of the executed felon' (Sawday, 1995: 153). In the dance also there is a lack of a sense of being in this body, it appears heavy and lifeless, lacking sensitivity as it flops between the three men manipulating it, characterizing it as object and 'flesh made word'.

However, the *Anatomy Lesson* presents another view of the body, also developed in *Blind Faith*. The painting portrays a theatrical performance acted out for a paying audience where the surgeons are also paid to 'star'. This visual theatre harks back to 'the Jacobean spectacle of the ... body *in extremis* ... to the overt, celebratory bodiliness of the dramatic and penal scaffold' (ibid: 65) also evident in the French post-structuralist Michel Foucault's discussion of the seventeenth-century 'Spectacle of the Scaffold', where he argues 'the condemned man published his crime and the justice that had been meted out to him by bearing them physically on his body' (1977: 43) The body in the painting is that of a thief, hence the dissection of the arm – the offending member. The visual theatre of the *Anatomy Lesson* is evident in the body undoing itself, holding open the folds of flesh in a highly Baroque way, derived from Vesalius's drawings, which reveal the unfurling layers of tissue and muscle down to the bone. The washing, examination, manipulation and dancing of the 'corpse' in *Blind Faith* are also visually theatrical. These theatrical bodies refer back to earlier notions of the body at one with its soul as part of the medieval universe, when 'subject and object are united through shared *meanings*, rather than rendered ontologically separate' (Bordo, 1987: 69). Rembrandt's painting and the dance seem to be representing *both* a body at one with its soul *and* a Cartesian soulless body textualized as an object of science. This is another instance of the Deleuzian notion that sameness and difference can occur simultaneously in one entity; that body and soul are both one and the same while also being different. The connectivity of body and soul and the accommodation of two opposing positions, are anti-Cartesian challenges to 'masculinized thought' which insists that 'each "sphere" remain distinct and undiluted by the other' (ibid: 114).

Fleshy corporeality is further foregrounded in *Blind Faith* when the dancers manipulate each other's near naked bodies on the table. Everyone is lifted, carried, turned around and over by their partners, in a more mutually supportive manner than earlier. This multiple animation, often using movement derived from contact improvization, emphasizes the materiality of corporeality by repetition. Everybody becomes both subject and object in this fluid folding of flesh on flesh, which results in a continual blurring of boundaries, as dancers move back and forth between active and passive roles, and pieta-like tableaux emerge and dissolve (Plate 12). The flesh in these ensemble scenes from *Blind Faith* is no longer inert matter subjected to the objectivity of science or matter expelled that needs to be restrained and kept in order. The machinations and closeness of these mutually supportive bodies seem rather to be celebrating the fluidity and diversity of fleshy bodies, embodying a different kind of subjectivity that could be associated with the feminine, with '"sympathy"... closeness, connectedness and empathy' rather than detachment and distance generally held to be masculine qualities (ibid: 112).

A focus on the textural characteristics of flesh in all three dances has strong affinities with Deleuze's descriptions of the form and texture of folds, which he terms 'the theater of matter'. This is characterized by the expressive features of matter created by light, depth, movement and 'the projection of something spiritual into matter' (1993: 37). The subtle rippling of muscles in Brown's back in *If You Couldn't See Me* creating concavities and convexities gives her back textural form. Movement of one muscle or bone affects others, the back's many different parts all connect to form a whole, which is its form, its 'theater of matter'. Deleuze describes Baroque folds, stating: 'texture does not depend on the parts themselves but on strata that determine its "cohesion"' (ibid: 37) and 'the object... is inseparable from the different layers that are dilating, like so many occasions for meanders and detours' (ibid: 37). 'Meanders and detours' suggest terrains and topography that can be mapped.[4] The topography of Brown's back in *If You Couldn't See Me* has been described as 'highlands' (Paxton, 1995: 94) and 'natural pathways' (Perron, 1996: 751), and its choreography as 'the workings of the body as remapped by Brown' (ibid). The close-ups of 'Joan's' face and ear starkly lit, and filmed in positive and negative, and the under-table lighting, combined with blue and gold washes of light on the near naked bodies in *Blind Faith*, also draw attention to the *materiality* of the flesh. The marble-like qualities of veined flesh are highlighted and the muscles stand out. In *Joan* a series of shots of tree branches lit and silhouetted against a black sky, and of the texture of internal amoeba-like and veined matter magnified under a microscope, are interspersed with 'Joan's' upturned head gazing into a blinding light, and with her kneeling folded body. Deleuze compares folds to 'veins in marble' (1993: 4) and he writes of 'an organic body' conferring 'an interior on matter' which is individual like the 'leaves of a tree' never alike 'because of their veins or folds' (ibid: 8). This focus on the textural qualities of the flesh in the dances further underlines the reinstatement of corporeality denied by the Cartesian 'masculinization of thought', allowing for the possibility of new subjectivities more associated with the feminine.

Fluids

Fluids, fluidity and patterns of flow are evident in all three works. Trisha Brown's movement style is often characterized as 'fluid'. Fluidity or flow is her hallmark. One critic has identified an 'incessant wave-like and spiral flow' as 'Brown's signature' (Ginot, 1997: 22), reminiscent of the wave-like characteristics of Deleuze's folds. Another writes that *If You Couldn't See Me* consists of 'fluid, limpid phrases' (Sulcas, 1995b: 38). It is a prime example of Brown's fluid style, with its sways and undulations of the torso, and swings, bends, lunges and rotations of limbs. Brown's body flows through positions and waves ripple through her pelvis. The choreography defies containment as parts of her body rarely seem to be in place or control. She seems to exceed and overflow boundaries, opening up possibilities for a new kind of fluid subjectivity. Since Brown never shows her face, an air of mystery surrounds her identity. There are parallels with *Blind Faith's* mysterious characters and the fluctuating identities of the mythical 'Joan'.

In *Joan* and *Blind Faith* internal bodily fluids are expelled in performance: 'Joan's' tears, and Clayden's saliva in *Blind Faith*. These fluids draw attention to the body's boundaries' permeability and the difficulty of distinguishing between inside and outside (Grosz, 1994b). Being neither inside nor outside, fluids are in between, and attest to the body's inability to contain itself. The expulsion of bodily fluids, which Kristeva terms abject, 'demonstrates the impossibility of clear-cut borders, lines of demarcation, divisions between the clean and the unclean, the proper and the improper, order and disorder' (Gross, 1990: 89). The abject 'disturbs identity, system, order' (Kristeva, 1982: 4). The subject's reactions to abjection 'represent a body in revolt' (Gross, 1990: 89). The social imposition of boundaries that limits and orders identity is derived from Cartesian philosophy, and its disruption, according to Bordo, can be linked to ideas associated with the feminine and with less fixed identities. This is what is happening in the dances, the abject is articulated in bodily fluids and 'dead bodies' – for Kristeva, 'the corpse is the utmost of abjection' (1982: 4). By articulating rather than expelling the abject, the dances open up new possibilities for subjectivity.

Other fluids feature in *Blind Faith*; the water used to wash Trigg's foot, later coloured red, and glasses of 'red wine' consumed by the dancers in the *The Last Supper* tableaux. When Christ instructs his disciples to consume bread and wine representing his body and blood, the word (of God/Christ) has been 'made flesh'. The tableaux of *The Last Supper* in

Blind Faith, provide references to the Passion where the word was made flesh. In Renaissance ideology 'the body of the world and that of the text are frequently identified with each other' (Barker, 1995: 21). They are considered at one 'in the figure of the Passion, where the word and the body are ... identified in an act of punishment and signification from which all other meanings flow' (ibid). This return to the word becoming flesh in *Blind Faith* further emphasizes corporeality and the permeability of bodily boundaries. The blurring of subject and object, self and other and mortality and immortality are symbolized in the Eucharist. In Dreyer's film, Joan is told that if she signs the confession her reward will be the body and blood of Christ and she is shown the Eucharist. The Eucharist's power resides in its association with Christ's Passion and ultimate immortality symbolizing the ambiguity of subjectivity – the possibility of life and death, God and man, subject and object, inside and outside existing simultaneously in one entity (see Kristeva, 1982: 118–20). Deleuze in his first book on cinema says of Dreyer's film that Joan's trial 'is itself Passion' which 'enters into a virtual conjunction with that of Christ' (1992: 108).

Kristeva's theories of abjection draw on these aspects of religious discourse. She indicates in her analysis of Holbein's *The Body of the Dead Christ in the Tomb* (1521) that 'because Christianity set that rupture [the splitting of the subject] at the very heart of the absolute subject – Christ; because it represented it as a Passion ... it brought to consciousness the essential dramas that are internal to the becoming of each and every subject' (Kristeva, 1989: 132). At the centre of Kristeva's theories is the notion of an unstable subject *in process* that is the antithesis of a Cartesian fixed subject. Kristeva sees 'holiness' as a domain where there is 'an excessive uncontrolled *jouissance* of ... transgressive ecstasy' (Grosz, 1989: 52).[5] She claims: 'the mystic's familiarity with abjection is a fount of infinite jouissance' (Kristeva, 1982: 127) and that 'we cannot escape the dramatic convulsions of religious crises' (ibid: 209). The transgressive ecstasy and dramatic convulsions evident in *Joan* and *Blind Faith*, in a Kristevan sense, could be said to be foregrounding the subject in process, thus departing from ideas of a fixed, unified, Cartesian subject. For Kristeva, 'abjection' is caused by 'what does not respect borders, positions, rules. The in-between, the ambiguous' (ibid: 4). This is connected with the feminine because, for Kristeva, the sacred or religious text is the 'revelation' of the feminine *chora* (Grosz, 1989: 84).[6] Religion displaces the abject associated with the feminine (Kristeva, 1982: 127). It makes space for the abject, allowing it to exist through transgressive ecstasy. There is an 'unrepresented residue ... that refuses to conform

... to masculine ... phallic order' which is 'occasionally touched upon by discourses of the sacred and ... experienced as religious ecstasy ... bliss ... surrender of a most corporeal kind' (Grosz, 1989: 84). The links between abjection and religious ecstasy involving 'surrender of a most corporeal kind' evident in Kristeva's theories are also evident in the dances. Commenting on Kristeva's statements about religious ecstasy, Grosz argues 'it is no surprise that saints' [she cites St Joan] 'could ... be defined as hysterics' (ibid). The expressions of religious ecstasy in *Joan* and *Blind Faith*, read through Kristeva's theories, can be seen to be bound up with the abject, and to be making way for new conceptions of an ambiguous subject in process, that draws on aspects of the feminine and departs from the rational, unified subject of Cartesian dualism.

Folds

Deleuze's theories of folds, developed from Leibniz's philosophy and read through ideas about the Baroque, refer to thought, philosophy and ideas, while also being evident in, or applicable to, a wide range of matter. Thus Deleuze writes of 'ideas' being 'folded in the soul' 'just as things themselves are inextricably wrapped up in nature' (Deleuze, 1993: 49). The examples he gives of folds in nature include 'the body ... waters, earth' and 'fabrics, living tissues, the brain' (ibid: 34), hence the relevance for the dances, where folded bodies are choreographed.

Connections between matter and thought in Deleuze's philosophy are also apparent in Brown's choreographic approach. 'Thinking about movement is, for ... Brown ... a sensually-charged alternation between moving thoughts and thoughtful movements' (Ginot, 1997: 20). Like Deleuze, her ideas seem to continually move between matter and thought; the connections between the two and the way the composition and behaviour of one informs that of the other. Brown plans her work in cycles, one of which she has termed 'unstable molecular structure'. '"Unstable" in that Brown constantly puts ... movement to the proof.... "Molecular" reminds ... that ... metamorphoses arise from a physical thought process that incessantly makes demands' (ibid). Deleuze uses 'molecular' and 'molar' to describe the composition of bodies and matter, where molecular suggests an oozing, laval, non-linear and unstable composition and behaviour similar to Brown's choreography. Deleuze and Guattari, describing music, assert, 'the molecular has the capacity to make the *elementary* communicate with the cosmic ... because it effects a dissolution of form' (1988: 308–9). There is a sense in which Brown's folding choreography for her back also effects a 'dissolution of form' that makes the 'elementary communicate with the cosmic'. Paxton mentions shifting 'from the specificity of ... visual solids to more oceanic senses' (1995: 94) and certainly when watching the ripples, stretches and undulations of Brown's back continuously, the patterns that emanate tend to waver between the visual and the visionary. The fold is characterized by connectivity both within and between folds. 'The fold always refers to other folds' and 'the organism is defined by its ability to fold its own parts ... one within another ... like Russian dolls' (Deleuze, 1993: 8). The connectivity that folds imply challenges notions of containment and unity traditionally associated with the masculine Cartesian subject. Deleuze indicates that this is the value of folds, stating: 'included in the category of things folded are ... philosophies that resolve Cartesian distinctions of mind and body through ... foldings ... that lead the eye to confuse different orders of space and surface' (Conley in Deleuze, 1993: xii). Different orders of space and surface are confused in all three dances where the fluidity of folds, implying movement, energy and multiplicity, is evident.

This kind of fluidity of a continually folding body is perhaps most evident in *If You Couldn't See Me*. Brown has been described in this solo as folding 'into the ground like an accordion' (Felciano, 1996: 11) and Wendy Perron, an ex-dancer of Brown's, writes of the dance, 'we see her finding new seams to fold on' (1996: 51). Deleuze describes the fluid characteristics of the Baroque as 'spongy, cavernous shapes ... put in motion by ... turbulence, which ends ... in the manner of ... a wave; matter seems to spill over into space' (1993: 4). Brown's dancing is often described as wave-like, as in 'the imperceptible wave forms, the inversions and vertiginous convolutions of Brown's movement ... [result in] a truly unstable structure' (Ginot, 1997: 20). In *Joan* and *Blind Faith* the matter of dancers' bodies 'spills over into space' via spilt bodily fluids and in *Blind Faith* in the ensemble choreography based on contact improvization.

The Deleuzian principle, derived from the fold, that sameness and difference can occur simultaneously in one entity is illustrated in Brown's work. Banes has said that her dances present 'the human body as both subject and object of research' (1980: 84). This ability to embrace sameness and difference simultaneously is demonstrated in Brown's extension of her solo *If You Couldn't See Me* into a duet for herself and a male partner. There is sameness (two dancers perform the same choreography simultaneously) and difference (facing different directions, one female, one male, and when Jones performs, one black body, one white body). The Deleuzian principle is underlined. Two dancers facing different directions demonstrate that things look different from different perspectives. Brown has said of the decision to make the duet with Jones 'it's important to embrace opposites' (1996) and she has talked about counterpoint in her work in terms of establishing the 'Other' (in Boxberger, 1997: 25). She seems to be well aware of the potential of relationships with 'otherness' or 'opposites' to affect ideas about identity. She and Jones have been described in their duet as 'unstable molecules' (Wesemann, 1995: 46), a reference perhaps to Brown's unstable molecular structures, but also possibly to the instability of identity (see Chapter 7).

The Deleuzian principle that sameness and difference can occur simultaneously is also apparent in the three dances because they are simultaneously visual and non-visual. The visual theatre style of the works enhanced by lighting, tableaux and foregrounding flesh is apparent, however, paradoxically the focus on corporeality, particularly its materiality, points to a bodily perception that is also distinctly *non-visual* (see Chapter 10). Flesh, fluids and folds are visceral, as the textural features of flesh indicate. The importance of corporeal perception as a way of knowing is implicated in the title *Blind Faith* which suggests a dependence on senses other than the visual. Bordo links the development of perspective, a visual appreciation of the world from a fixed point, which contributes to a notion of 'geometrical seeing' (1987: 67), with the development of Cartesian philosophy. Both can be seen to be part of the 'masculinization of thought' she identifies, because both rely on a separation of subject and the world – a sense of distance and detachment (see Chapter 10). Snaith's title for her dance and the emphasis throughout on bodily senses other than the visual, are evidence of a search for a different kind of subjectivity no longer determined by Cartesian philosophy.

Brown turning her back to spectators is another example that disrupts the logic of visualization. There is a long tradition in her work of challenging spectators' perception, disrupting objectification, by having dancers on rooftops, walking down walls and floating on rafts, for example (see Chapter 3). Brown subverts 'the optical logic and location of spectatorship' (Lepecki, 1997: 16). By turning her back to the audience in *If You Couldn't See Me*, Paxton claims, she 'refutes . . . frontal convention' (1995: 94). He continues: 'facing up relieves you of facing us . . . you aren't blinking uncomfortably in the light of our avid eye. You . . . cannot know or concern yourself with how you may look to us... we are not watchers.... You are not our focus.' (ibid) Objectification of Brown's female dancing body becomes difficult, if not impossible. According to one critic: 'the work is... devoid of... that eroticism... associated with women putting themselves in front of the public' (Felciano, 1996: 11). The importance of the non-visual, is also apparent in *Joan*, through 'Joan' closing or covering her eyes, and as in *If You Couldn't See Me*, 'Joan' turns away from her audience, so that only the back or top of her head is seen. These departures from the logic of visualization in the three dances are all evidence of a flight from the Cartesian masculinization of thought and moves towards a more embodied, feminine kind of subjectivity.

Ideas of multiplicity, openness and excess in the Baroque and folds are evident in the dances where excessive and multiple dancing bodies open up possibilities of new subjectivities. Deleuze's multiplicity importantly acts on borders and boundaries, he writes of: 'the Baroque... dividing divergences into as many worlds as possible... by making... many possible borders between worlds' (1993: 81). Openness, excess and multiplicity when explored alongside borders suggest a merging of inside and outside which is evident in the dances. Deleuze argues, 'the outside and the inside... are... not two worlds' (ibid: 31) and referring to modern art, 'Stockhausen's musical habitat or Dubuffet's plastic habitat do not allow the differences of inside and outside... to survive' (ibid: 137).

Brown's layered choreography has affinities with Deleuze's notions of multiplicity and layering implied in folds. By opening up the choreographic structure of her solo in the duet, *You can see us*, Brown is exposing how it works. She has done this before, for example, in *Opal Loop* (1980) two identical duets are performed simultaneously; one with the couple together, and the other with them split apart. Brown's strategies of accumulation that layer choreographic material in the manner A, AB, ABC, and so on, are well-known (in for example *Accumulation*, 1972, and *Split Solo*, 1974). Describing her work, the dance critic Rita Feliciano writes, 'sometimes... there is so much going on within a single body that it seems impossible to take it all in' (1996: 7). 'Little folds... unravel in every direction, folds in folds, over folds, following folds' (Deleuze, 1993: 86), and 'physical mechanisms... work by... communication and propagation of movement, "like ripples that a stone creates when it is thrown into water"' (ibid: 97). Deleuze's statements resonate with Brown's choreography. Deleuze writes of the force and energy of folds, stating, 'movement... cannot be stopped' (ibid: 12), 'the fold ... moves... between essences and existences. It... billows between

... body and soul' (ibid: 120). Links between body and soul are continually suggested in *Joan* and *Blind Faith*. When 'Joan' looks upwards towards light, and is diffused with light, her body *becomes* light, evoking a connectivity between body and soul. This imagery is inspired by Dreyer's film, which Carney suggests 'link[s] bodies and spirits and urge[s] their unity' (1989: 255). Deleuze also writes of the spirituality in Dreyer's film stating, 'Dreyer produces the triumph of a... spiritual perspective' (1992: 107).

The energy of folds to which Deleuze refers when writing of body/soul relations is also apparent in the continuous folding of bodies and 'corpses' in *Blind Faith*, particularly with its references to Baroque and Renaissance paintings and sculptures. These references are given prominence through the device of tableaux, where bodies assemble in enfolded masses and dissolve only to fold themselves into other tableaux (Plate 12). The concept of tableau, based on the belief that a single instant within a narrative was all that could be portrayed within painting, was dominant in art theory from the mid-sixteenth to the mid-eighteenth century. The tableau enhances and singles out everything in its field. 'Everything that it admits... is promoted into essence, into light, into view... the tableau is intellectual, it has something to say... it knows how this must be done' (Barthes, 1977: 70). The tableaux in *Blind Faith* show bodies draped over each other like folds of cloth, limbs intertwining, shoulders, arms, heads cascading down from above. They resemble characteristically Baroque works such as Tintoretto's *Paradise* (1588), where piles of bodies and limbs coalesce in an infinitely intricate pattern animated by light falling on folds of flesh. In tableau 'an ideal meaning' (ibid: 74) is 'communicated at a glance' (Burgin, 1986: 88), embodying 'understanding... wisdom and substance' (ibid: 90) conveyed through the body and gesture. The art theorist Victor Burgin suggests that corporeality is often elevated to a position of mystical significance in tableaux, which have been compared with hieroglyphs which 'stand outside discourse' (ibid) and involve 'a meaning which will not be pinned down by words' (ibid). The tableaux in *Blind Faith* involve distinctly kinesthetic meanings, enhanced by the movement of the dancers from one tableau to the next, which transcend words. The 'word' becomes 'flesh' further foregrounding corporeality within the dance.

The mystical dimensions of the Baroque that permeate Deleuze's writing on folds have resonances with mystical elements in the dances. 'The age of the "Baroque Gothic" witnessed the birth of the mystical experience... characterized... by an individual's account of his or her voyage to and from an ineffably universal event, which set the body in a trance' (Conley in Deleuze, 1993: xii). *Joan* and *Blind Faith* portray spiritual quests which 'set the body in a trance'. 'Joan's' relationship with her god is a spiritual journey culminating in her wandering through woods, with flames at her wrists and her tunic hem, looking around, as if searching, and then closing her eyes, appearing mystical and trance-like as she transcends the flames, unhurt. In *Blind Faith* Snaith acts as a catalyst investing the others with energy, light and life, exploding the notion of a modern, thinking, bodyless subject by evoking ecstatic, trance-like mystical performances. *Blind Faith* seems to encapsulate a spiritual journey. Its Programme note states: 'the work is structured in three distinct phases, beginning with "darkness and delusion", moving through "death and transformation" and opening out into "light and levitation"' (1998).

The excessive trance-like dancing, which evokes the Baroque, has links with abjection in Kristeva's theories. She also makes references to the Baroque, stating, 'all art is a kind of counter-reformation an accepted Baroqueness' (Kristeva, 1987: 253). For her, both art and religion have excessive Baroque-like qualities in the form of 'transgressive eruptions' that enable them to challenge the symbolic (Grosz, 1989: 53). Kristeva also associates a 'new baroquism' with an 'ambiguisation of identities' or 'the fact that people don't have fixed identities' (Kristeva, 1984: 23). The non-fixity of identity which I argue is evident in the dances through the foregrounding of corporeality, was also a concern of Dreyer, whose characters have been described as liberating themselves from 'fixity, coherence and stability' in their roles as 'figures ... in continuous movement and redefinition' (Carney, 1989: 96).

Mysticism and spirituality have also been associated with Brown's solo where she has been described as dancing with 'an almost spiritual intention' (Wesemann, 1995: 46). Paxton explores these qualities in more depth when he writes, 'you are ... a medium mediating between us and ... some unknowable or unthinkable vision' (1995: 94). Deleuze, writing of folds in painting, claims, 'in every instance folds of clothing acquire an autonomy and a fullness.... They convey the intensity of a spiritual force exerted on the body ... to turn it inside out and to mold its inner surfaces' (1993: 122). This force turning bodies inside out is apparent in the dances. The folds of cloth and flesh in the intertwined postures and gestures of the disciples in *The Last Supper* are animated in *Blind Faith*. The 'disciples' appear to be enfolded together through the power of the Holy Spirit entering into them at that moment.

Conclusion

Reading *If You Couldn't See Me, Joan* and *Blind Faith* in the light of the theories of Deleuze, Bordo and Kristeva, has demonstrated the ways in which imagery in the dances can be seen to have connotations concerning the fluidity of subjectivity. The materiality of bodies, evident in flesh, fluids and folds, emphasized through references to painting, sculpture and film, and through the powerful mode of tableau, has foregrounded the blurred inside/outside spaces of bodies resonant with new possibilities for subjectivity. In these in-between spaces bodies are no longer seen as containers with separate insides and outsides, bodies and souls, but as fluid transformative entities disrupting hierarchical binaries bound up with masculine ideas of a Cartesian subject. Flesh, fluids and folds of bodies have been seen to constitute, shift and transgress bodily borderlines. By focusing on bodily flesh, folding bodies in close contact, excessive ecstatic bodies, fluid movement, abject bodily fluids and corpses, the dances emphasize body/space interfaces, where negotiations take place concerning the redefinition of identity. As the British feminist philosophical theorist Christine Battersby observes: 'Identity as understood in the history of Western philosophy since Plato has been constructed on a model that privileges . . . self-contained unity and solids' 'what is missing from our culture is an alternative tradition of thinking identity . . . based on fluidity and flow' (1993: 34). Exploring the inside/outside spaces of the dances through a Deleuzian reading of the Baroque, in conjunction with Bordo's gendering of Cartesian philosophy, and Kristeva's theories of abjection, has revealed ways of thinking identity based on fluidity and flow. The dances reinstate the materiality of corporeality and its connectivity with spirituality, associated with a more embodied and fluid feminine notion of subjectivity.

In the next chapter bodies and their excesses remain the focus, but the emphasis shifts from the sacred body out of control to excessive 'grotesque' bodies associated with the carnivalesque, and the ways in which they subversively trouble fixed notions of subjectivity associated with a contained body.

9
'Carnivalesque' Subversions in Mark Morris' *Dogtown*, Liz Aggiss' *Grotesque Dancer* and Emilyn Claid's *Across Your Heart*

Introduction

Whereas the previous chapter focused on the mystical qualities of folded and fluid flesh, this one considers the grotesque potential of transgressive bodies in *Dogtown* (Morris, 1983), *Grotesque Dancer* (Aggiss and Cowie, 1987, revived 1998) and *Across Your Heart* (Claid, 1997). These dances, although quite different from each other, share carnivalesque grotesque qualities. On one level they seem either funny or disturbing, or perhaps both, but I argue that they go further than this. Using the Russian literary critic Mikhail Bakhtin's theories of the carnivalesque, I show that these works, by transgressing and exceeding the traditional inside/outside boundaries of the body, can be subversive and suggest ways of rethinking subjectivity.

Bakhtin developed his theories mainly through his analysis of Rabelais' (1494–1553) writing, evident in *Rabelais and His World* (submitted as a thesis in 1940, first published in 1965, and translated into English in 1968). Bakhtin explores the ways in which the spirit of carnival activities based in the folk tradition of popular culture, both literally and conceptually can be transposed into art, and the relations between art and the surrounding popular cultural forms. He sees the annual Lenten carnival as a historical phenomenon in early modern Europe and as a mode of critique of officialdom. Key features of the carnivalesque that he explores include the grotesque body, laughter, marketplace speech, banquet imagery and the lower bodily stratum. From carnival Bakhtin derives the notion of 'grotesque realism' to describe humanity's shared experience of a basic material body that is open to the world. Bakhtin's work was in part a prescriptive and utopian model of socialist collectivity. His theories of language and literature which can be applied to other arts, stress the multiple meanings in texts, the instability of the sign as referent and intertextual relations between texts, all of which anticipate post-structuralist developments.

The focus of Bakhtin's theories of the carnivalesque on the body and performance make them particularly valuable for dance. Bakhtin comments that in Rabelais' work 'images of the human body' play 'a predominant role' and that 'similar traits' found in 'other representatives of Renaissance literature in Boccacio, Shakespeare, and Cervantes, were interpreted as a "rehabilitation of the flesh" characteristic of the Renaissance' (Bakhtin, 1984:18). Ideas associated with this same Renaissance 'rehabilitation of the flesh' are explored in the previous chapter. Bakhtin derives this bodily imagery from 'that peculiar aesthetic concept... characteristic of this folk culture... grotesque realism' (ibid). Bakhtin argues that 'all... forms of grotesque realism degrade, bring down to earth, turn their subject into flesh' (ibid: 20). Degradation and debasement of higher forms are inherently subversive. Carnivalesque performances and grotesque bodies degrade officialdom, turning themselves and it inside out and upside down. Their unruliness threatens the order, stability, hierarchy and control of a world of binary oppositions.[1] The body's role is central because the transgressions of boundaries that Bakhtin identifies 'are effected through... the body' (Stallybrass and White, 1997: 301). According to Bakhtin, the ambivalent space of carnival creates a topsy turvy world of parody and play where traditional boundaries between performers and audiences are eroded. The intermingling of performers and spectators in carnival provides 'interaction', 'interchange' and 'interorientation' with the world. Although these dances exist within the logic of the spectacle of theatrical performance and not in a carnival context, Bakhtin's category of the carnivalesque bridges the extension of carnival categories for use as more general social critique. In this sense, despite Bakhtin's theories predating postmodernism, they are pertinent for these readings of postmodern dance since, their emphasis on the materiality of the body and its interorientation with the world place them alongside other antidualistic theories and point to possibilities for a rethought embodied subjectivity. Through an analysis of the grotesque body, they show how subjectivity can be fluid, ambiguous, multiple and marginal rather than fixed, and through a focus on bodily imagery and carnivalesque performance, they can point to underlying subversive elements in dance that might otherwise go unnoticed.

The utopian nature of Bakhtin's theories and his claims for the radical and political powers of carnival have prompted criticism. The extent to which the carnivalesque can be a force for change, given that it is a licensed and permissable outlet for feelings authorized by 'official culture' to occur on certain named feast days, has been questioned (Eagleton, 1981). Another criticism is that 'a history of actual carnival reveals that the marginalized – Jews, women, homosexuals – could become the victims of ritual punishment' (Morris, 1994: 22). However, as cultural theorists Peter Stallybrass and Allon White point out, if the carnivalesque is recognized as 'an instance of a wider phenomenon of transgression' it is possible to 'move beyond Bakhtin's... *folkloric* approach to a political anthropology of *binary extremism* in class society' (emphasis in original) (1997: 301). They argue that this goes 'beyond the... debate over whether carnivals are politically progressive or conservative' revealing 'that the underlying structural features of carnival operate far beyond the strict confines of popular festivity and are intrinsic to the dialectics of social classification as such' (ibid). In other words the carnivalesque and the grotesque have extended beyond historical categories to become epistemological terms.

The dances

Grotesque Dancer is a 30-minute solo created by the British artists, Liz Aggiss and Billy Cowie, performed by Aggiss. It is loosely based on the avant-garde dances of the twenties and thirties' German cabaret dancer, Valeska Gert, whose distorted movements portraying low-life city characters such as prostitutes, were frequently described as 'grotesque'. The music for *Grotesque Dancer*, composed and performed by Cowie, is a series of songs and instrumental numbers with texts by Goethe, Morgenstern and Dehmal arranged for voice, piano and saxophone. For the 1987 version, the music was on tape, but for the revival it was performed live on stage. The music style is expressionistic with the vocal part veering from lyrical melody, through speech to *sprechgesang*.[2] *Grotesque Dancer* consists of a series of cabaret numbers or vignettes – five dances or mimes, and five or six songs depending on the version. Aggiss performs in a single spotlight on a small stage and she sings into a microphone on a stand. Her performance is extremely intense (Plate 13). It includes bouncing up and down rhythmically to a German folksong, stretching as if training for sport, inching across the stage robotically in a pastiche of a marionette, singing passionately into the microphone while gesticulating to the audience and twisting into angular, angst-ridden postures. The stark follow spot, her make-up – bright red lipstick and heavily made-up eyes – and her costumes – black satin knickerbockers and white top with white knee-length socks and black flat shoes for the first half, and long, black, satin, evening gown and stiletto heels for the second half – emphasize the theatricality of the piece. Various theatrical shocks are provided by Aggiss' mood, character and costume changes; one minute she appears young and eager, the next coy and teasing, the next haggardly and harsh. The piece's excessive contorted movements, its parodic style, transgression of gender boundaries and plays with performer/audience relationships suggest associations with the grotesque and the carnivalesque.

Gender and animal/human boundaries are played with in American Mark Morris' **Dogtown** (1983), an 11-minute dance for two men and five women from the Mark Morris Dance Company, to five songs by Yoko Ono, including the eponymous 'Dogtown'. The seven performers, wearing different coloured bras and panties or trunks over lycra shorts, execute sequences of repetitive movement often in unison in a 'doggy' style, on all fours, frequently in lines. The performers simulate urination and copulation foregrounding inside/outside body interfaces. The focus on base behaviour of the lower bodily stratum is a characteristic of Bakhtin's grotesque body evident in the excesses of *Dogtown*. The set consists of piles of full plastic bags evoking urban rubbish. The choreography ridicules human, urban sexual behaviour, poking fun at the habits and rituals of supposedly sophisticated inhabitants of modern towns and cities. There are associations with Bakhtin's descriptions of the bestial behaviour at early European carnivals and fairs and of the frank and free atmosphere of the medieval marketplace.

Across Your Heart, created by the British choreographer, Emilyn Claid, is a 70-minute work performed by CandoCo, some of whose dancers have disabilities. To commissioned music by Stuart Jones, it was created for four women (Helen Baggett, Celeste Dandeker, Charlotte Derbyshire and Sue Smith) and three men (Jon French, Kwesi Johnson and Kuldip Singh-Barmi); two of the performers, Dandeker and French, are in wheelchairs. Theories of grotesque bodies and the carnivalesque can be particularly pertinent when considering the ways in which bodies with disabilities are sometimes viewed and treated. Their inclusion in dance, where bodies are typically and traditionally constructed as classical, bounded and sleek and this is the norm, as Albright (1997) suggests, can often be disruptive and subversive in a grotesque and carnivalesque manner. *Across Your Heart*, episodic in structure, includes scenes derived from sexual, religious and other fantasies with bawdy, overtly sexual dancing, singing, a wheelchair parade (Plate 14) and a South American carnival. The piece employs parody and humour which is sometimes degrading and perverse. It begins and ends with Kuldip Singh-Barmi as a distorted, Calibanesque figure in a clinging, long dress (just visible in the top right corner of Plate 14) performing angular, twitchy, hunched, grotesque movements and Helen Baggett as a female Christ-like figure in a white loin cloth initially standing on a platform facing away from the audience. Gothic horror movies were one source and they are evidence of a postmodern blurring of high and low culture in the work.[3]

In this chapter, challenges to the binaries associated with gender, sexuality and ability in the dances are examined in the light of Bakhtin's characterization of the grotesque body as open and permeable and the carnivalesque as ambivalent. Excesses are then investigated in the performance styles of *Grotesque Dancer* and *Dogtown* and in suggestions of leaky bodies in *Dogtown* and *Across Your Heart*. Links with the French-based Bulgarian feminist Julia Kristeva's theories of abjection, explored in Chapter 8, are identified. An examination of the ways in which carnivalesque parody in the dances can degrade official discourse follows, exploring Bakhtin's theories concerning the role of humour and the regenerative power of laughter. Finally, grotesque and carnivalesque interactivity with the world perceived in focuses on the lower body, the downward thrust of movement towards the earth and a propensity for floorbound movement in the dances is explored, and linked with Bakhtin's suggestions of the 'down to earth' elements of folk traditions and popular culture. Throughout I identify the ways in which Bakhtin's theories reveal the dances' subversive potential, associated with inverting hierarchies and their connectivity with the world suggestive of renewal.

Grotesque challenges to boundaries

The openness of the grotesque body and the permeability of its boundaries are seen by Bakhtin as bringing the body of the people closer to the world. This is graphically illustrated in Bakhtin's analysis of the story of Gargantua's birth in Rabelais' novel which occurs at a 'merry banquet' during the 'feast of cattle slaughter' before Lent (Bakhtin, 1984: 220). Gargamelle, who gives birth, has consumed an excess of tripe (intestines) from fattened cows at the feast, despite warnings from her husband that 'there are no intestines without dung' (ibid: 223). As a result, her right intestine falls out which is of an 'unsavoury odor' and it is initially mistaken by midwives for the birth. Bakhtin suggests 'Gargamelle's labor and the falling out of the... intestine link the devoured tripe with those who devour them. The links between animal flesh and the consuming human flesh are dimmed, very nearly erased. One dense bodily atmosphere is created' (ibid: 221–2). Bakhtin argues that the carnivalesque atmosphere permeating the episode 'ties into one grotesque knot the slaughter, the dismemberment and disemboweling, bodily life, abundance, fat, the banquet, merry improprieties and finally childbirth' (ibid: 222). Gargantua is born through his mother's ear, which as Bakhtin comments, 'is a typical carnivalesque turnover' 'the child does not go down but up'(ibid: 226). Employing this imagery in part for social commentary, Bakhtin concludes that 'we see looming beyond Gargamelle's womb the . . . womb of the earth and the ever-regenerated body of the people' (ibid). Thus grotesque bodily imagery with permeable boundaries brings the 'body of the people' closer to a regenerating earth.

Bakhtin claimed that fifteenth- and sixteenth-century literature of the kind that Rabelais and others produced, pointed to a world where hierarchies are diminished and inverted; Gargantua's birth through the ear is a physical example and a parody of the religious doctrine of virgin birth. He saw the blurring of the grotesque body's boundaries, evident in the example of Gargamelle's consumption of tripe and the intestines' expulsion blurring binary oppositions such as self/other that are part of the construction of a unified, rational subject, separate from the world, also discussed in Chapter 8. Bakhtin argues, 'the grotesque ignores the impenetrable surface that closes and limits the body as a separate and completed phenomenon. The grotesque image displays . . . the outward and inward features . . . merged into one' (ibid: 318). This is because the gut as the lining of the intestines, where food passes having entered through the mouth to exit through the anus, is both inside the body but also a continuation of the outside. In this sense Bakhtin suggests that 'grotesque imagery constructs what we might call a double body' (ibid).

This concept of a 'double body' can inform a reading of Liz Aggiss' performance in *Grotesque Dancer*. In her black satin knickerbockers, a chest flattening white top and boyish haircut, her performance of repetitive physical jerks – knee bends, stretches and push-ups – executed strenuously, emphasizing the musculature in her arms and legs, is decidedly masculine. When she later dons her black satin evening gown and gold stiletto heeled shoes her appearance is transformed from masculine to feminine. Her dance is interspersed with singing at a microphone, in the style of male and female cabaret singers. Throughout *Grotesque Dancer* her 'double body' transgresses the gender binary, her performance plays between seductive femininity and aggressive masculinity. One reviewer described her as 'part Liza Minelli and part football player' (Constanti, 1987: 26). She also looks androgynous, her maleness and femaleness blur, in the grotesque style of her performance. After five numbers, when donning the satin dress, she shockingly rips off a wig to reveal a shaved head (Plate 13). The combination of her baldness with harsh make-up and long black satin gown is uncanny and startling, at least in part because of its androgyny. For an audience not knowing whether a performer is male or female can be unsettling as indicated in relation to *Out on the Windy Beach* in Chapter 4.

Aggiss' double body also flouts the binary between strength and vulnerability. In the physical exercises and stunts of her performance Aggiss demonstrates her strength. One routine has her hopping on one leg for several minutes while contorting herself stretching. We spectators marvel at her ability to keep her balance while bouncing up and down in time to the insistent rhythm of the music. Aggiss' powerful singing at the microphone projecting directly to her audience, extended by amplification such that her voice is everywhere, is further evidence of her command. Yet alongside these confident displays of dominance there are touching moments of vulnerability. Towards the end of the piece she is reduced to crawling across the floor, satin gown in mouth, and finally, looking disillusioned with the façade of performing, she drops her gown to the floor, steps out of one shoe and limps off stage. Aggiss has stated in interview that in order to achieve the combination of strength and vulnerability required in performance she had to find two sides in her character and alternate between them (in Briginshaw, 1988: 11). Her grotesque performance and this insight into the resources required for it, point to a double body of the kind Bakhtin describes, that transgresses boundaries and binaries in its openness.

Aggiss' relationship with her audience in performance contains an ambivalence that resonates with Bakhtin's descriptions of the blending of self with the world in carnival. The intermingling of performers and audience typical of carnival is evident in cabaret/night club performances. *Grotesque Dancer* was commissioned by the Zap Club in Brighton, a small venue situated in a cellar-like space, for which Aggiss and Cowie created the piece. In interview Cowie characterizes the cabaret-like performance of *Grotesque Dancer* as 'the idea of not being so much up on a pedestal but actually being in touch with the audience in a club'. Aggiss adds, 'we could see how it would work in that kind of smokey, small place, inviting the audience to be right on top of it, inviting them to practically participate' (ibid: 9). Aggiss' performance includes much direct eye contact with the audience, particularly when she is singing, and when she is dancing and displaying her strength in various physical feats, she raises her eyebrows as if to say 'look what I can do!'. The twinkle in her eye suggests she is making fun of her own performance self-reflexively. This gives *Grotesque Dancer* a worldly quality. In Bakhtin's terms 'the confines between . . . the body and the world are overcome: there is an interchange and an interorientation' (1984: 317). Aggiss is sharing her performance with her audience knowingly, she is 'one with the crowd . . . not . . . its opponent . . . [s]he laughs with it' (ibid: 167). The ambivalent merging of self and world, self and other, performer and audience is a blurring of binaries that disrupts the power relationships inherent in oppositions of this kind. As Bakhtin points out, 'Rabelais continually used the traditional folklore method of contrast, the "inside out", the "positive negation". He made the top and the bottom change places [and] intentionally mixed the hierarchical levels' (ibid: 403). Aggiss achieves similar objectives in *Grotesque Dancer*. By mixing gender values and norms in her grotesque performance she achieves an ambivalent androgyny, making 'the top and the bottom change places'. Her mixture of strength with vulnerability is a kind of 'positive negation'. Her performance challenges boundaries and binaries, going beyond hierarchical norms and values associated with a closed, unified subjectivity, suggesting instead a more multiple, open subject connoted by her 'double body'.

Carnivalesque's ability to turn things upside down and inside out by blurring boundaries and disrupting binaries is evident in Mark Morris' *Dogtown* where the boundaries of gender and sexuality are transgressed. The piece opens with three couples facing front, with partners one behind the other, the rear dancers' arms are clasped around the chests of their partners. Rhythmically miming Ono's lyrics: 'let me take my scarf off, no, no, no, don't help me' the dancers in front jerkily pull back their partners' arms in an attempt to free themselves, but they are quickly grasped again. Those behind overpower those in front. Different actions are repeated to different verses of the song. To the words of the next verse 'let me take my blouse off', the hands of the rear dancers grab the breasts of the dancers in front who pick them off finger by finger in a grotesque manner, yanking them away before being gripped again. The partnerships are differently gendered. Two are male/female; one with the woman in front, the other reversed, and one consists of two women. The performance is almost unpleasantly sexual. Given the actions and words of the song, it appears as if the rear performer is attempting to forcefully strip their partner against their will. Expectations that the male is dominant in a sexual relationship and that sexual liaisons are predominantly heterosexual are challenged by this performance. Bakhtin comments that in 'typical carnival. Everything . . . is inverted in relation to the outside world' (ibid: 383). By transgressing the norms of gender and sexuality in a carnivalesque manner Morris is challenging his audience to think beyond the limits of conventional expectations of gendered and sexual behaviour.

Inversions that challenge binaries of sexuality and ability in a grotesque and carnivalesque manner are evident in *Across Your Heart*. In one scene three women, wearing black and decorated in silver chains, like a gang of s/m lesbians, strut their stuff in a confident streetwise manner. Approaching a statuesque woman in a long white dress, who has just been seen bride-like on the arm of a man, they taunt her, fingering the fabric of her dress. They flaunt themselves in front of her clenching their fists, squatting and sauntering past her, running their hands through their hair. This bodily 'mockery and abuse' is 'grotesque' in Bakhtin's terms (ibid: 319). Their poses appear deliberately performed to create an effect. These stereotypical caricatures of s/m lesbians become increasingly threatening as they execute a dynamic unison routine of kicks and pivoting turns. Their performance and costumes assert the difference between them and the 'innocent bride'. The trio rock 'the bride' from side to side and grotesquely mimic her puppet-like floppy arm movements. They then remove her dress revealing a black-clad body encased in a steel frame which enables this paralyzed woman to stand.

The tension of difference is replaced with a sisterly solidarity of sameness as the metallic support and black garb eroticize the former bride's body in a queer way, similar to the chains on the others' bodies. Now when the threesome tip her from side to side, lift her and lower her, there is a gay abandon that was not there before. She smiles with them while embracing them in turn and laughs her enjoyment; she has joined the club. According to Bakhtin, 'terror is conquered by laughter', and the carnivalesque by bringing the world closer 'freed man from fear', turning the world into 'a sequence of gay transformations and renewals'(ibid: 394). This scene in *Across Your Heart* presents 'gay transformations and renewals' in a carnivalesque manner.[4]

The exposure of the disabled grotesque body supported by a metal frame also refers intertextually to the Mexican artist, Frida Kahlo, who repeatedly painted self-portraits revealing her grotesque and freakish body, held together by a metal spinal insertion (see, for example *The Broken Column* 1944). Claid has played with the meaning of 'grotesque' by representing the norm in the form of the bride-like, Kahlo look-alike, feminine figure. Juxtaposed initially with the three chain-clad women in black, she positions them as deviating from the white and pure norm she represents. They appear grotesque, but because of her stilted, limited movement, and the intertextual references to the Mexican Kahlo, in a way so is she, and certainly when she is disrobed and revealed as a 'freak', who can only stand in a metal support, she joins them. In joining them she has crossed the binary and demonstrated the fluidity of notions of beauty and ugliness and of the carnivalesque power of the grotesque. Like *Grotesque Dancer* and *Dogtown, Across Your Heart* has blurred binaries; those of sexuality and ability. The performance is disruptive and unsettling, it points to the instability of the signifiers of sexuality and ability. The monstrous bodies of the 'lesbians' and the 'bride', through their juxtaposition and choreography, exceed their boundaries, challenging notions of fixed identities that exist either side of a binary. They have opened up possibilities for a new communal relationship within a larger queer 'body of the people' in Bakhtin's sense.

Excessive overflows

'Exaggeration, hyperbolism, excessiveness are generally considered fundamental attributes of the grotesque style' (Bakhtin, 1984: 303). The openness and permeability of the grotesque body often result in excessive overflows such as the expulsion of bodily fluids. Bakhtin claims: 'the grotesque is interested only in that which . . . seeks to go out beyond the body's confines' (ibid: 316). Kristeva's theories of abjection, discussed in the previous chapter, also concern expulsion of material elements of corporeality considered improper and unclean, such as bodily fluids and waste. The similarities are not surprising. Kristeva was one of the first to introduce Bakhtin's work to France. For her, his work importantly developed a dynamic model of literary texts, which he saw as open to contestation and unresolved, or as 'carnivalesque riots' (Hill in Fletcher and Benjamin, 1990: 143). Kristeva adopted this emphasis and developed it within a psychoanalytic framework for her concept of the subject in process. Affinities with the open, unfinished nature of Bakhtin's grotesque body are evident. The abject, like the grotesque body and aspects of the carnivalesque, threatens to overflow and disrupt the order, containment and control of the rational, unified, closed subject.

Overflow of bodily fluids is hinted at in *Dogtown* when the dancers mimic dogs urinating and copulating, less obviously they wear surgeons' thin rubber gloves. Given their minimal costumes, the addition of rubber gloves seems bizarre. They could relate to protection from the 'urban rubbish' surrounding the dancers. However, given all the simulated sexual activity in the dance, and the time of its creation –1983 – when AIDS had just come to public attention as a subject of concern in the West, and was regarded predominantly as a sexually transmitted disease of the gay community, they could also be a reference to one of the most visible reactions to the disease; wearing rubber gloves to protect against contact with infected body fluids. Morris is openly gay and has referred to AIDS in interview indicating that his life expectancy could be considerably limited (Acocella, 1993: 115). The rubber gloves are a sign of control of the abject. Having copulating doggy creatures wear them however, dispenses with any serious notion of control, it is a bizarre carnivalesque joke. It mocks their normal use and is perhaps a wry comment on official culture's reactions to AIDS. Bakhtin, referring to sexually transmitted diseases, writes 'gout and syphilis are "gay diseases", the result of overindulgence in food, drink, and sexual intercourse. They are essentially connected with the material bodily lower stratum ... [and] widespread in grotesque realism' (1984: 161).[5] The baseness and nearness to the earth of the 'lower bodily stratum', a much cited feature of the grotesque and carnivalesque body, for Bakhtin is an indication of its positive interconnectivity with the world (see final section of this chapter).

This interconnectivity with the world is evident in the atmosphere of the medieval marketplace, associated with the carnivalesque and the grotesque, some features of which are apparent in *Dogtown*. The dancers spend much of their time crawling on all fours like dogs among bags of rubbish, which they add to by simulating urination. The lyrics of Ono's *Dogtown* which accompany the performance include cries of 'pease porridge hot, pease porridge cold, some stays in the pot nine years old'. This nursery rhyme is 'the rhythmic beat of the vendor' that Bakhtin identifies as 'steeped in the atmosphere of the market' (ibid: 170) and 'advertising' food. Carnival excess is often expressed in feasting. Bakhtin argues: 'the banquet images – food, drink, swallowing' 'aspire to abundance' 'they rise, grow, swell ... until they reach exaggerated dimensions' (ibid: 278). Pease pudding is a messy viscous substance that could be likened to loose excrement, it also encourages flatulence. Bakhtin claims that 'the images of faeces and urine are ambivalent, as are all the images of the material bodily lower stratum; they debase, destroy, regenerate and renew simultaneously' (ibid: 151). Kristeva's concept of the abject, which includes such substances, is similarly ambivalent because it exists both inside and outside the body blurring its boundaries. *Dogtown's* deliberate foregrounding of the base life of contemporary city dwellers steeped in imagery of waste in general and bodily waste and fluids in particular, challenges portrayals of the human subject as controlled, fixed and stable. It also suggests that the performers in *Dogtown* have sunk to the depths of the city – the sewers – which Stallybrass and White have argued are 'low and grotesque', they are the city's 'lower bodily stratum'(1986: 143, 145). By opening up the body's boundaries *Dogtown* challenges restrictive binaries, and associations with the lower bodily stratum portrayed, in Bakhtin's terms, ambivalently connote regeneration and rebirth suggesting possibilities of renewal and change.

At the end of *Across Your Heart*, Helen Baggett, dressed as a Christ-like figure is seen hanging upside down over Celeste Dandeker who sits at the head of a banquet with four of the other dancers. Singh-Barmi, costumed in his Calibanesque garb, holds a wine glass under the hanging body as if to catch blood. The glass is filled with a red liquid which he drinks and then passes to Dandeker who also drinks as the lights fade signalling the end of the dance. This act has associations with the bloodthirsty vampire and horror movies which inspired *Across Your Heart*. Consumption of the blood of the body of the Christ-like figure also has connotations with the Eucharist, but this body is female and hanging upside down in a doubly blasphemous, carnivalesque fashion. Bakhtin cites 'oaths and profanities' as key elements of marketplace and carnivalesque speech which 'were mostly concerned with sacred themes' for example, '"the blood of Christ"' (1984: 188). Bakhtin saw these profanities, as 'the unofficial elements of speech . . . as a breach of the established norms of verbal address; they refuse to conform to conventions . . . etiquette, civility, respectability. These elements of freedom . . . create a special collectivity, a group of people initiated in familiar intercourse, who are frank and free in expressing themselves' (ibid: 187–8). For Bakhtin these carnivalesque utterances were positive challenges to the language of officialdom, freeing it up and in the process creating a social 'body of people' with the potential for challenging the limitations of symbolic discourse which operates through binary oppositions. *Across Your Heart's* shocking ending challenges the fixity of gender by having a female Christ, while foregrounding the abject in a doubly grotesque manner by having her co-performers drink her blood. For Kristeva, as outlined in the last chapter, when the abject is revealed rather than hidden, the binary coding of the human subject in terms of inside/outside and self/other is challenged pointing to the ambiguity of the subject, to an open or unfinished subject, which has parallels with Bakhtin's grotesque body. This unfinished subject in process, because of its openness, has the facility for connection with others and with the world in a liberatory and renewing manner, according to Bakhtin. *Across Your Heart's* powerful, ambivalent ending by challenging binaries that separate and value male over female and sacred over profane, provocatively suggests the potential for rethinking subjectivity in this way.

Other excessive overflows are evident in the performance styles of *Grotesque Dancer* and *Dogtown*, where the dancers perform often in an almost robot-like fashion, driven by the incessant beat of the accompanying music. Bakhtin writes of the 'mighty torrent' of the grotesque body flowing through Rabelais' 'entire novel' (ibid: 323). The commitment and relentless, repetitive, almost mechanical movements in these dances also appear as an unstoppable 'torrent', an excessive outpouring of bodily energy, almost like freakish acts at the fair. Aggiss' virtuoso feat of balance and stamina in her 'hopping dance' has already been mentioned. In *Dogtown* one phrase of movements with the dancers lined up, lying side by side, consists of alternate dancers rolling one way while the others 'jump' themselves stretched out on hands and feet over the dancer next to them who is rolling underneath them. Immediately they land they take their partner's place lying down supine to be jumped over. This miraculous feat of continuous strenuous activity is performed repeatedly. The performers in *Grotesque Dancer* and *Dogtown* seem to exceed their bodies in this grotesque repetition. As one reviewer asserts: 'what is more grotesque than the movement itself is Aggiss' incapacity to stop it' (Constanti, 1987: 26).

Just as the rubber gloves in *Dogtown* control the 'abject', wheelchairs can be seen to control and contain the 'unfinished', 'incomplete' 'grotesque' bodies of the dancers with disabilities. When bodies with disabilities leave the containment of their wheelchairs in a sense they are going beyond themselves in an excessive outpouring or overflow. As Bakhtin argues 'the grotesque image . . . retains . . . that which leads beyond the body's limited space' (1984: 317–18). In *Across Your Heart* Dandeker is seen out of her wheelchair twice; in her metal frame during the Kahlo scene described earlier, and when she is seated regally on a large sumptuous bed-like structure and attended by three of the men. Jon French is tipped out of his chair by Singh-Barmi and Kwesi Johnson, who, after taunting him, wrestle with him on the ground and then wheel the chair off leaving him apparently 'helpless'. Charlotte Derbyshire appears with the wheelchair and tries to help French back into it, but she is told by French in no uncertain terms to leave him alone initially. She can be seen as representative of official culture, in Bakhtin's terms, trying to maintain order and control. French's rebuff makes it clear that although he is dependent, he will not be controlled. After the parade when all the dancers except Singh-Barmi are in wheelchairs (Plate 14), several move in, out of and between the chairs upsetting any sense of uniformity. The messiness of these bodies that exceed their wheelchairs goes against the traditional aesthetic of controlled, sleek, effortless, 'classical' bodies silently executing 'beautiful' moves in more conventional theatrical dance. It highlights the contrast between what Bakhtin terms 'the new bodily canon' – the 'finished, completed, strictly limited body. That which protrudes, bulges, sprouts or branches off . . . is eliminated, hidden or moderated. All orifices of the body are closed' (ibid: 320) – and the grotesque. Bodies that do not fit the Western norm by being overweight or having disabilities can be seen as 'unfinished', 'protruding' and 'grotesque' in Bakhtin's terms. But Bakhtin's reappropriation of the grotesque interprets it as becoming and regenerative rather than lacking. *Across Your Heart's* excessive overflows, in terms of bodies with disabilities that exceed 'normal' boundaries can be seen as grotesque in Bakhtin's terms and hence regenerative. By exceeding the limitations of the norm, evident in closed, contained, silent, able bodies, they suggest new ways of perceiving disability specifically and physicality generally.

Carnivalesque parody

Much of the renewable force that Bakhtin claims comes from the grotesque and carnival is, he suggests, generated by the power and energy of laughter. Humour is the lifeblood of carnival and the medieval marketplace. Its power comes from its ability to turn things upside down and to mock, evident in parodic acts and plays with language. Key contributions come from puns which open language up to the possibility of double meaning and the performances of clowns who mimic. Through language and performance, gaps can be opened in the symbolic, otherwise closed, official discourse of the day. These plays and parodies can allow and encourage critiques of the oppressive regime of the time. In this sense they reveal positive possibilities for renewal and change in Bakhtin's terms. Bakhtin describes medieval laughter as 'the social consciousness of all the people' experienced in the marketplaces and carnival crowds when 'man' 'comes into contact with other bodies of varying age and social caste' as 'a member of a continually growing and renewed people' (ibid: 92). He claims that laughter presents 'victory ... over supernatural awe, over the sacred, over death; it also means the defeat of power, of earthly kings, of the earthly upper classes, of all that oppresses and restricts' (ibid).

Dogtown is a parody of human urban behaviour and sexual promiscuity. The automaton-like movements of the performers in straight lines, following one another on all fours, makes them look as if they are programmed. Their close-fitting, colourful, brief, lycra costumes, possibly refer to similar attire worn to aerobics classes and fitness sessions at city gyms, but the dancers' bodies do not match the lythe, sleek ideal associated with such activities and their performance is mockingly comic and grotesque. Their very funny simulated doggy behaviour includes 'urinating' and 'copulating' in time to the musical beat. Guillermo Resto and Tina Fehlandt alternately rear back and up on their 'hind legs' lifting their rubber gloved hands in the air, one above the other, paw-like, and mime growling. Yet because of the look and costuming of the performers, the double meanings of the piece are only too apparent. It is a playful, carnivalesque, fun-poking dance, reflecting the mood of Ono's ironic songs. Through parody it deconstructs and challenges the norm of the glossy veneer of one version of contemporary city life.

Grotesque Dancer is also a parody, mimicking and mocking the genre of female performance where women are on display for the spectators' pleasure. One reviewer writes, 'as Aggiss bats her legs in a parody of high kicks ... a shocking contrast between aesthetic beauty and unglamourised physicality is exposed' (Farman, 1987: 20). Valeska Gert, the inspiration for Aggiss' performance, created her deliberately grotesque shows for Dadaist venues where she mocked typical Berlin cabaret acts displaying women as sexual objects (see Jelavich, 1993). As well as 'grotesque', her performances have been described as 'comic' and 'savage clowning', including 'satirical imitations of existing dance styles' (De Keersmaeker, 1981: 60). These elements of satire and parody are also evident in Aggiss' performance, which shifts between different theatrical modes from singing to dancing to posing in the spotlight, foregrounding the frame of performance. This is a performance about performance. One reviewer writes, '*Grotesque Dancer* is heavily stylised ... conjuring up images of ... clowns' (Farman, 1987: 20). Bakhtin sees the medieval clown as a key figure of carnivals and fairs who, through travesty and parody, is able to turn things inside out and upside down. Aggiss' performance, through exaggerating typical postures and gestures of cabaret, mocks its contortions and feats, deconstructing it. This is the kind of performance that, as the American queer theorist Judith Butler argues, is '*theatrical* to the extent that it *mimes and renders hyperbolic* the discursive convention that it also *reverses*' (her emphasis)(1993: 232). When Aggiss limps off at the end, leaving her dress and one shoe behind, she breaks the performance frame revealing the underlying effort, usually hidden. The ridiculous and bizarre lengths women are expected to go to as entertainers are exposed in *Grotesque Dancer*. This grotesque display challenges the norms and expectations of female performance. Aggiss stated in interview, '*Grotesque Dancer* threw up quite a lot of contentious issues because it didn't fit into the normal pattern of dance . . . it created a lot of real antagonism . . . it . . . opened up quite a lot of things for people which is great. I mean change is due – all the time really' (in Briginshaw, 1988: 11). The ways in which the piece is contentious, opening up issues and bringing about change, parallel the ways in which Bakhtin claims carnivalesque parody works to challenge power and bring about change.

Across Your Heart in its plays with some of the darker aspects of the human psyche inspired by sexual fantasies, horror movies, religion and death has resonances with the parody of the church and the power of ecclesiastical dogma that Bakhtin outlines. The blood drinking scene, involving a female, Christ-like figure, has already been mentioned. In another scene four dancers are gathered round a 'dead body' laid out on a plinth, placing flowers on it. The sombre mood gradually changes as one of the dancers begins to giggle and laughter infects the other three. They begin to play with the body as if it were a life-size male doll. They sit him up, open his eyes, move his mouth and arm getting him to prod one of them with his finger. One of them tests the body's reflexes and a knee shoots up. All four dancers laugh hysterically. They stand the body up, one of the women dances with it and sits it down on another's lap, she parts its legs with hers. This macabre mockery of the supposedly serious subject of death generates something similar to the 'remarkable symphony of laughter' Bakhtin writes of in Rabelais' work (1984: 163). Laughter is a constant theme in Bakhtin's writing, particularly its power to overcome fears such as the fear of death. He states, 'folk culture strove to defeat through laughter this extreme projection of gloomy seriousness and to transform it into a gay carnival monster' (ibid: 395).

The eruption of laughter around the dead body in *Across Your Heart* is in keeping with other carnivalesque elements in the piece. At one point a carnival float emerges to the sounds of gay Latin American music. French sits on the float in his wheelchair wearing large feathered wings, the other performers all in carnival masks wheel the float forward, they dance around it lewdly gyrating their pelvises and feeling their crotches. They lick each other's fingers, kiss and bite each other, lie across French's lap, handle breasts and bums and simulate buggery. This carnivalesque performance, in its focus on bodily orifices and on base sexually and orally gratifying activity, includes many grotesque elements that Bakhtin identifies. In another scene the company stage a wheelchair parade with all seven dancers in wheelchairs creating synchronized designs (Plate 14). A reviewer writes, it is a parody of 'wheelchair formation dancing' (Howard, 1998: 315). Throughout *Across Your Heart* irreverent and unruly behaviour, of the kind Bakhtin describes as grotesque, dominates. Discussing the role of the image of the Bakhtinian grotesque body in the work of sixties' artists in Greenwich Village, the dance theorist Sally Banes argues, 'it is . . . by means of the image of this grotesque body of misrule that unofficial culture has poked holes in the decorum and hegemony of official culture' (1993: 192). By including the raw physicality of the grotesque body in carnivalesque scenes in *Across Your Heart* including dancers with disabilities, Claid has exploded the expectations and norms that often delimit and contain bodies with disabilities. In this sense the 'misrule' of the 'unofficial culture' of *Across Your Heart* has 'poked holes in the decorum and hegemony of official culture'.

Grotesque and carnivalesque interactivity with the world

For Bakhtin one of the positive features of the carnivalesque is the closeness to the earth often suggested in its imagery. He values this because of its regenerative traits – the earth is a source of renewal and of rebirth – and because of associations with interconnectivity with the mass of the people. For Bakhtin, the lower bodily stratum also symbolizes regenerative power since it is where birth takes place and its lowness renders it near to the earth. Regenerative traits of the grotesque and carnivalesque are identified by Bakhtin in frequent references to the 'lower bodily stratum', in the downward thrust and low baseness of much grotesque movement and in the propensity for grotesque bodies to link with other bodies.

All three dances include much downwardly directed movement of the sort Bakhtin describes frequently as 'downward thrusting'. He claims:

> The mighty thrust downward into the bowels of the earth . . . is directed towards the underworld . . . This downward movement is

... inherent in all forms of popular–festive merriment and grotesque realism. Down, inside out ... upside down, such is the direction of ... these movements ... all that is sacred and exalted is rethought on the level of the material bodily stratum ... the accent ... is on the descent ... downward movements ... are ... understood anew and merged in one ... movement directed into the depths of the earth, the depths of the body.

(1984: 371).

In *Grotesque Dancer* Aggiss plunges her fists down between her open knees as if into the depths of her body. A central duet in *Dogtown* begins with Tina Fehlandt's arched torso descending dramatically to the ground. The fight that results in *Across Your Heart*, when French is tipped out of his wheelchair by Singh-Barmi and Johnson, occurs on the ground as the three men wrestle and roll over each other, evoking the 'downward movements' of the medieval 'fights, beatings and blows', that Bakhtin cites (ibid: 370). When Dandeker is out of her wheelchair in her metal walking frame, she is only able to make movements that go down heavily to the earth. All these examples invert the elevation of classical dance with its emphasis on lightness, uplift and flight associated with idealism, in favour of the materialism of the earth. Although much modern dance is also contrasted with classical dance because of its earthbound movements and use of gravity, the context and meanings associated with the downward movements of modern dance I would argue are very different. They are bound up with a modernist search for and expression of a single 'self', 'meaning' 'truth' or 'origin' that postmodern works no longer recognize.

Emphasis on the lower bodily stratum and movement near the ground is evident in *Dogtown* where dancers frequently 'mount' each other, sniff each other's behinds and gesture to their own behinds. Much of the time the performers are on the floor. They crawl on all fours, roll over each other, lie curled up on the ground and lie down with their legs open. They also throw themselves and others to the ground. There is a vivid moment when one of the women carrying a man wrapped round her waist flings him to the floor, this is repeated three times. This emphasizes the earthbound 'low life' of *Dogtown* while also transgressing 'normal' gender boundaries. Morris has said of his company 'my guys are articulate and my gals are brutish. And they're both both. They can all do everything' (In Acocella, 1993: 91).

The exaggeratedly grotesque performance style of *Grotesque Dancer* often emphasizes the lower part of the body. Aggiss is frequently on the ground lying or sitting with her legs open, at times she appears to tie herself in knots, her legs appear over her head as she rolls over her shoulders. In a physical exercise routine she bends over with her back to the audience, her knees bend and her behind rapidly bounces up and down in time to the music. While doing this she looks through her legs and first one hand is passed through and grabs the opposite calf, then the other follows. Next she clasps her thighs in a similar fashion, lastly she places her hands on opposite cheeks of her bottom. This contorted posture is both gross and freakish. Bakhtin quoting from Schneegans' *History of Grotesque Satire* (1894) describes an incident cited from the Italian *commedia dell'arte* where Harlequin butts a stutterer in the abdomen to assist him to get a word out. Schneegans likens this to childbirth. Bakhtin claims,

> a highly spiritual act is degraded and uncrowned by ... transfer to the material bodily level of childbirth ... thanks to degradation the word is renewed ... reborn ... we ... see the ... bodily hierarchy turned upside down; the lower stratum replaces the upper stratum. The word ... localized in the mouth and ... head (thought) ... is transferred to the abdomen and ... pushed out ... Here ... we have the logic of opposites, the contact of the upper and the lower level.
> (1984: 309)

This, Bakhtin sees as a paradigm example of the way in which the grotesque can invert. The body since the Renaissance has been constructed as a vertical hierarchy, the lower parts of the abdomen and genital area are associated with baseness, and the head, with higher, more valued ideals linked with the mind. This hierarchical and polite bodily self-presentation is codified in ballet (see Chapter 10). Bringing the head down to, and in Aggiss' performance, below, the level of the genitals and behind, challenges these assumptions based on the body/mind binary, and visually turns them upside down.

In *Across your Heart* there are also moments when heads come close to genital areas but in a much more overtly sexual manner. While Baggett, in a glittering sequined bikini, sings into a microphone, two other women perform pelvic gyrations directly in front of the faces of two men in her 'audience', the women begin to pull down their trousers for the men revealing glittery underpants beneath. This erotic performance in the dingy half-light suggests a seamy night club cabaret act. It provides another instance of Bakhtin's grotesque 'logic of opposites, the contact of the upper and the lower level' (ibid).

When the dancers are seen on and around the carnival float in *Across Your Heart* a certain interactivity with the world in Bakhtin's terms is evident. They appear as a moving tableau kissing and clasping each other, often lewdly. They move up, down and over each other's bodies, wheelchairs, and the float, in monstrous couplings. At times, inside and outside body surfaces merge to form one 'body of the people'. Bakhtin writes: 'the pressing throng, the physical contact of bodies, acquires a certain meaning. The individual feels that he is an indissoluble part of the collectivity, a member of the people's mass body . . . people become aware of their sensual, material bodily unity and community' (ibid: 255). In this scene from *Across Your Heart* differences of gender and ability dissolve into the mass of this becoming body of the people.

Conclusion

Exploring Bakhtin's theories alongside these three dances has foregrounded their subversive and oppositional elements. The acceptable witty side of carnival is immediately evident in the dances but this has a tendency to obscure the darker, messier, elements that can challenge authority and shock. Applying Bakhtin's theories has enabled readings of the dances that not only reveal these shocking elements but also provide a way of understanding their significance. Aggiss' sudden revelation of her bald pate, hands grasping breasts in *Dogtown* and blood drinking in *Across Your Heart* have each in different ways been seen to be subversive, regenerative and suggestive of new meanings. Binary oppositions involving gender, sexuality, ability and religion have been challenged. The power of parody to pervert and invert, and the regenerative qual-ities of humour and laughter, have been demonstrated. Although the dances remain within a theatre art context and space, the links with popular and folk culture of not only carnival but also cabaret, horror movies and nursery rhymes in the dances, together with the excessive use of the lower body and downward movements, through a Bakhtinian reading have been seen to invert values in a provocative and subversive manner. The value of Bakhtin's grotesque conception of the body is that 'it "degrades" the human form in a positive way . . . bringing its subjects down to earth, it re-embodies what official culture had disembodied or etherealized' (Banes, 1993: 193).

This chapter has brought together American and British examples of postmodern dances which, each in their own way, blur 'high' and 'low' or popular culture. A Bakhtinian reading aids an understanding of why all three dances reject the aesthetic of grace embodied in classical ballet and some modern dance. In a piece such as *Dogtown* for example, the awkward, ungraceful movement style of this relatively early Morris work is not immediately noticeable because of the precision and rhythmical clarity of the performance. Applying Bakhtin's notions of the grotesque body and carnivalesque excess has revealed more subversive elements allowing alternative readings. By transgressing the inside/outside boundaries of bodies both physically and conceptually in a wide range of often shocking and provocative ways, the three dances have demonstrated a subversive re-embodiment in their grotesque and carnivalesque features. Notions of a fixed, stable, complete subject that is not open to change have been problematized. In their place the dances have suggested ideas of an unfinished and becoming subject in process, that is open and ambivalent with permeable boundaries, allowing for excessive overflows with the potential for renewal, and for forgetting the self in favour of submerging and merging with the 'body of the people'.

In the final chapter the focuses of each of the three parts of the book, namely constructions of space and subjectivity, in particular site specific places, negotiations in actual and conceptual in-between spaces and the folds and excessive flows of inside/outside body/space borderlines, are shown to overlap in the readings of dances that engage with and construct architectural space. The chapter explores how these dances challenge the logic of visualization, bound up with the single viewpoint of perspective of the rational unified subject and, through different ways of experiencing space, suggest other possible subjectivities.

10
Architectural Spaces in the Choreography of William Forsythe and De Keersmaeker's *Rosas Danst Rosas*

Introduction

Throughout this book body/space interfaces in dances have been examined and discussed focusing on the particular locations of site specific works, on actual and metaphorical in-between spaces and on inside/outside bodily borders in order to rethink identity and subjectivity. This chapter brings these ideas together by examining dances which play with inside/outside architectural spaces in a deconstructive manner. I am using 'architectural space' here to mean spaces that are structured actually or conceptually according to ideas associated with building design. William Forsythe's choreography creates architectural spaces, and the filmed version of Anne Teresa De Keersmaeker's *Rosas Danst Rosas* (1983, film version 1997), is set in and interacts with the architectural spaces of an empty school building (Plate 15). In Forsythe's *Enemy in the Figure* (1989) the stage space is divided by a wavy wooden screen and various lighting effects into architectural spaces, like rooms, which the dancers move between (Plate 16). In this sense the dancers are seen inside and outside different spaces. In *Rosas Danst Rosas* dancers are filmed inside and outside the school building and its rooms where the dance is set. They are filmed from within, from without and through glass windows and doors rendering the space of the building at times difficult to fathom.

Both dances are about deconstructing space, about defamiliarizing space such that it is experienced differently. Deconstruction is concerned, through dislocation and defamiliarization, to expose the gaps and to reveal the free play of meanings inherent in texts, in this instance the spatial texts of dance. It opens up and reveals the limits of things such that they cannot be put back together in the same way. It is unsettling. This is what happens to the space of the dances. It is deconstructed because the choreography and performance disrupt the 'logic of visualization' or the way in which space is traditionally seen and experienced. This experience of space is bound up with what it means to be a subject and with constructions of subjectivity. There are parallels in dance and architecture which reveal the pervasive ways in which ideas about experiencing space and the visualization of space shape thought and affect subjectivity. Dance and architecture as spatial texts structure ways of seeing the world. Dances and architecture can be seen to organize space. Predominantly they do this through a logic of visualization bound up with notions of perspective, associated with a masculine way of viewing the world derived from Cartesianism, which was discussed in some detail in Chapter 8 (see also Bordo, 1987). Seeing things from a particular perspective or viewpoint locates the viewer, affecting their sense of subjectivity. As I outlined in the Introduction, in this view a self or subject is constructed separate from the world and seen as rational and unified. As well as being gendered, these ideas, derived from a logic of visualization, are also 'racialized' and valued because of their associations with moral rightness.

Dances and architecture can disrupt this logic of visualization by blurring separations between the insides and outsides of bodies, buildings and space, and thus creating and working in in-between spaces. This is doubly evident in those dances set in and around buildings. Forsythe's choreography in general and De Keersmaeker's *Rosas Danst Rosas* in particular disrupt the logic of visualization by the ways in which they engage with architectural space. Forsythe and De Keersmaeker approach the experience of space and subjectivity in very different ways, but on a certain theoretical level they are both concerned with blurring these separations which affect the construction of subjectivity. They both refuse to present subjectivity in terms of straightforward representation and being. As a result, what they are doing spatially in their choreography does not make sense if viewed in a conventional way. Certain post-structuralist and deconstruction theories aid an understanding of the significance of their refusal to present representation and being in a conventional manner.

Representation is created through the logic of visualization. It is dependent on a particular way of seeing the world where what is created in terms of an art or dance work is believed to re-present some sort of metaphysical 'meaning' or 'truth', which can be discovered and shared by an audience. By disrupting the logic of visualization both Forsythe and De Keersmaeker refuse to engage with this notion of representation.

Classical ballet and modern dance in the ways they present the dancer and represent 'meaning' both address and position viewers as rational unified subjects. Classical ballet does this through its vocabulary and aesthetic which creates total, harmonious, balanced beings. Parallels can be seen in Ancient Greek Vitruvian architecture which is based on a classical, mathematical system of proportions believed to be the same as the ideal human form. The architect, Daniel Libeskind, who is often linked with Forsythe because they have worked together and share a similar approach to space, deconstructs this kind of classical architecture, just as Forsythe deconstructs classical ballet, refusing to look at bodies and space in a classical way. In this chapter insights from Libeskind's architectural practice are used to illuminate some of Forsythe's deconstructive strategies in dance. In modern dance the dancers appear as beings who are rational, unified subjects because the body is seen to express inner 'meanings' or emotions, in a way that it 'never lies'. In this sense the body is seen as transparent. De Keersmaeker's dancers resist this kind of transparency because of the ways in which they merge with the architecture and with each other in *Rosas Danst Rosas*. They do not appear as separate beings struggling to express their individuality.

Both Forsythe and De Keersmaeker, in the ways in which their dances engage with architectural space, are concerned with geometry, but their handling of geometry is very different. Where Forsythe is undoing the geometry of classical ballet, De Keersmaeker chooses to *use* geometry. Forsythe deconstructs classical geometric proportions, which relate to space and the human form, in order to find new ways of experiencing space beyond the linear and the visual, which open up possibilities for a different kind of unfixed subjectivity. De Keersmaeker seems to be doing the opposite, although the end result is similar. By accepting and creating geometrical grids in her choreography excessively, she denies her dancers an individual subjectivity. They become part of space through the repetitive geometrical patterns they perform, which deny them a sense of individuality. Both choreographers in their different engagements with architectural space and geometry disrupt the logic of visualization and point to new possibilities for subjectivity.

In this chapter I explore how perspective and the logic of visualization are bound up with particular constructions of the subject, and through examining these choreographers' works, show how dance, partly through certain parallels with architecture, can challenge and disrupt this logic and suggest alternative possibilities for subjectivity. After a brief introduction to the choreographers and dances, the ways in which the dances disrupt the traditional single viewpoint of perspective and challenge notions of a separated self, are examined. Next links between space and subjectivity are introduced. They are used to explore the role of the visual in the construction of bounded and gendered subjects and ways in which the dances challenge and disrupt this construction process are outlined. The dominance of the visual in traditional experiences of space and the ways in which the visual is bound up with reason, construction of a rational subject, and particular ideologies and values, are then outlined. Finally the ways in which the dances trouble these notions, by disrupting the visual and finding other non-linear, discontinuous ways of experiencing space are explored and discussed.

The choreographers and dances

William Forsythe, the American choreographer who directed Ballet Frankfurt from 1984 to 2004, and the Forsythe Company since then, is noted for the treatment of space in his choreography. He emphasizes a non-linear approach to space and choreography inspired by fractals and chaos theory. Mark Goulthorpe, an architect who has worked with Forsythe, has suggested that his work proliferates the 'rhizomatic experimentation' of the French post-structuralists Gilles Deleuze and Felix Guattari (1988) who also draw on chaos theory, and that Forsythe creates 'ballets of disorientation and trauma' (1998). I focus on Forsythe's 30-minute *Enemy in the Figure* (1989) with scenery, lights and costume designed by Forsythe, and refer to the dance video *Solo* (1995) directed by Thomas Lovell – a six-and-a-half-minute solo for Forsythe. Both have music by Thom Willems.

Enemy in the Figure, performed on its own, and as the central part of *Limb's Theorem* (1990), is distinctive for its lighting; a large lamp on wheels moved around by the dancers, and its set; a wavy wooden screen dividing the stage space diagonally from upstage right to downstage left and a length of thick rope on the floor, which is moved and undulated by dancers. The performance space is constructed and deconstructed throughout by the dividing screen and the moveable lamp (Plate 16). The piece opens with one dancer on her back in the upstage right corner by the length of rope illuminated by the lamp, the rest of the stage is in darkness. She is attended by another dancer crouching at her side moving her limbs. They seem to be experimenting with what limbs can do, stretching them, possibly warming up for a dance class. The lamp is wheeled downstage facing the back wall gradually illuminating more and more space. Other dancers become visible; overall there are six female and five male dancers. Throughout the dance, the lamp, which appears to be the only source of light, is moved about ten times, sometimes in darkness. Each time it illuminates a different part of the space, sometimes just showing a solo or duet or at others various groupings. Highly technically trained balletic bodies performing both non-balletic and balletic vocabulary such as attitudes, pirouettes and arabesques, but in an off-balance, off-centre, distorted and fractured manner, characterize these dances. Willems' electronic score of shattering metallic sounds often provides a regular beat which dancers pick up and drop seemingly at random. When they pick it up they occasionally slip into jazz and street dance forms such as breaking and body popping, which are frequently mixed with ballet vocabulary. Dancers walk purposefully across the space to begin their solo or duet, or to go and lean up against the wall and watch for a while. Every so often performers are seen running fast round the periphery of the stage emerging from behind the screen only to disappear again. 'Lots of things happen at once, so the eye is drawn hither and thither' (Gilbert, 1998: 13). 'The dancers' space (and the audience's perspective) is continually modified' (Sulcas 1991a: 32). A sense of chaos and discontinuity is evident.

Often the lamp illuminates one part of the stage and light spills over into other parts where dancers in semi-darkness, or shadows of activity can be glimpsed. What can be seen clearly and what is only in half-light, shadow, or behind the screen is in a continual state of flux. Seeing and not seeing, appearance and disappearance, are threads running through the piece, resulting in 'no . . . fixed spectatorial vantage points' (Brandstetter, 1998: 50). The dancers' black and white costumes enhance the sense of appearance/disappearance. The audience has to look actively to fathom what is going on where at any one point. The dance critic Rosalyn Sulcas says of *Limb's Theorem*, 'it is . . . [an] investigation and elaboration of the relationships between dance and its perception . . . light is deployed as a means of exploding and contracting the space, and of disturbing the spectator's visual certainties' (1991b: 37). Forsythe comments, 'It forces you to re-examine those things [that you take for granted, which perhaps you've stopped seeing] and say ". . .what are these things?"' (in Driver *et al.*, 1990b: 93). Forsythe is challenging spectators to depart from dependence on a single perspectival viewpoint that fixes and distances. He has stated, 'let us move into the text' (in Brandstetter, 1998: 52) and criticizing proscenium theatres, it would be better if audience and dancers could touch each other' (in Odenthal, 1994: 33).

Anne Teresa De Keersmaeker, the Belgian choreographer, who has directed her company *Rosas* since 1983, is based at La Monnaie, Belgium's national opera house, from where she also directs PARTS (Performing Arts Research and Training Studio), where training is conducted by ex-dancers of the choreography of William Forsythe, among others. She has stressed the significance of spatial composition and strategic uses of space in her work, stating 'space, volume, structure and trajectories are very important to my work, both choreographically and dramatically' (in Hughes, 1991: 17).

De Keersmaeker's ***Rosas Danst Rosas*** was originally choreographed for the stage in 1983 for four female dancers including De Keersmaeker, with music commissioned from Thierry de Mey and Peter Vermeersch. In 1997 de Mey made a 57-minute film version set in the RITO school in Leuven, designed by Belgian architect Henry van de Velde (1863–1957), beginning with four female dancers and ending with 18. The film opens with a shot of one dancer seen from above through the gridded, rain-splattered panes of a glass conservatory roof in the evening half-light. As the camera moves it is sometimes difficult to make her out. This sets the scene for what is to follow. The dancers merge with the architecture as they are frequently glimpsed through glass windows, doors and walls of the building and their grey school uniform-like skirts at times merge with the school's grey walls. The dancers, in precise geometrical formations, repeat highly punctuated, unison stepping, spinning, twisting and bending-over phrases with much energy and panache. The drill-like style of the performance seems relentless as the dancers' sighs and intakes of breath are heard and, like their movements, keep precise time with the percussive accompaniment. One long sequence is performed sitting on chairs where the dancers cross and uncross their legs, place their arms crossed over their knees and fling their heads forwards and backwards, their hair flying. Interspersed within these almost robotic patterns of repetitive movement are more everyday gestures, such as running hands through hair and pulling tee-shirts on and off the shoulder. Their obsessive repetition makes them seem neurotic as they are absorbed into the tight structure. The young women do not appear aware that they are being filmed; only occasionally do they look at the camera or each other. Relations between the architecture and the choreography are emphasized through linear repetition in both. Panning and moving cameras and quick edits further highlight relations between the two. The spatial characteristics of the distinctive location are foregrounded in the filming, when the dancers recede into the distance and play in different spaces of the school, between themselves, and themselves and the building, the camera, and their virtual spectators. Seeing the performers through windows and doorways from inside and outside blurs these boundaries, such that the space becomes fragmented and discontinuous; it is seen from many viewpoints rather than one.

Disrupting the single viewpoint of perspective

Arguments concerning the reductive, limiting and masculine nature of ideas about perspective and visualization have been rehearsed elsewhere in this book, in the Introduction and in Chapter 8. The idea of perspective, which emerged during the Renaissance but was developed from Ancient Greek Euclidean geometry, implies a single viewpoint in space from which and to which all points converge. As the feminist theorist Susan Bordo argues, 'the "point of view" of the perspective painting . . . spatially freezes perception, isolating one "moment" from what is normally experienced as part of a visual continuum' (1987: 64). Most dances and much architecture order space by isolating particular moments in this way. The world becomes a 'series of framed images . . . each one a particular "perspective"'(ibid: 65). Architecture and dance also order space when they act as 'architectural and cultural "cues" that teach us to "read" the world' as if it consists of fixed images seen from a single viewpoint, or 'Euclideanly' (ibid: 66). Quoting the cultural theorist Patrick Heelan, Bordo gives examples of these 'cues' as 'simple engineered forms of fixed markers, like buildings, equally spaced lamp posts and roads of constant width' (1987: 251). Space is not framed in this way in the work of Forsythe and De Keersmaeker. They both disrupt the single viewpoint of perspective by presenting dance as a spatial experience, as a visual continuum, which militates against the production of 'cues' or markers in the way Bordo describes.

Enemy in the Figure provides no cues as to how to read the space. Solos, duets and group dances do not clearly start and finish, they just appear and disappear; there are no single viewpoints, no images are framed and fixed. One clear perspective is replaced by many possible focuses of attention, rendering the work polycentric. When the moveable lamp is turned round and round the shadows it throws up multiply everything in view (Plate 16). The architect Steven Spier writing of Forsythe's work in the *Journal of Architecture*, says that it 'assumes many points and . . . axes' (1998: 138) and the dances 'tend to have multiple centers of interest . . . things happening extreme right or left, in front of the worst seats, and obscuring vision' (ibid: 140–1). He claims a Forsythe performance 'cannot be a simple whole with a clear center' (ibid: 140). Polycentricity, disorientation and a lack of frontality, are apparent in *Enemy in the Figure*, exacerbated by the angles of the set and the light which carve up the space. In other words, as the social theorist Paul Virilio has said about contemporary experiences of space, 'the substantial, homogenous space derived from classical Greek geometry gives way to an accidental, heterogeneous space in which sections and fractions become essential' (1991: 25). Polycentricity is evident in both the choreography and the performance of individual bodies. Forsythe uses disorienting strategies with dancers, having them work as if the front of the room has become the floor or a diagonal, or as if the floor is tilted (Driver *et al.*, 1990b: 92), so that they experience space differently. In *Enemy in the Figure* a woman appears to tie her body in knots as she travels low to the ground, limbs twisting and turning over and under each other, reaching out in all directions. With her torso twisted one way, her head looking another way, and legs stretched wide, her male partner, who can hardly be seen in the semi-dark with his black skin and dark suit, lifts her. She remains stiff like a sheet of contorted metal, until he lets her slide down his shins to the ground and she unfurls and is away. This fractured, sometimes jerky, sometimes fluid, discontinuous experience of space is typical of the piece.

Daniel Libeskind, the deconstructivist architect best known for designing Berlin's Jewish Museum, says of Forsythe's treatment of space: 'there are . . . particular diagonals which differentiate his view of space from the orthogonal spaces . . . so often the scenography of contemporary dance' (Libeskind, 1999b). Orthogonal spaces, made up of straight lines and right angles, are examples of the 'cues' for reading and ordering space in a perspectival sense. In Forsythe's work lines are exploded and broken. They no longer operate as cues in the sense Bordo describes. At one point in *Enemy in the Figure* about ten dancers, all facing different directions, perform jerky, robotic movements, some freeze for a moment in disjointed positions, others are spinning, others have limbs flailing in different directions, some are jumping at odd angles, others leap or run across the space. Any sense of pattern, line, order or coherence is exploded. There is no single perspective and no clues as to how to view this chaotic ever-changing mass of performers. As the dance theorists Patricia Baudoin and Heidi Gilpin identify, 'both Forsythe and Libeskind move toward . . . an opening of the . . . assumptions of their respective disciplines. In Libeskind's hands, linearity is lost when the architectural model is exploded: "The rational orderly grid actually turns out to be made up of a series of decentered spaces"' (1989: 20).

Where Forsythe decentres and fragments space, disorienting spectators and performers alike, De Keersmaeker reinforces the linearity and grid-like patterning of perspectival space. Where Forsythe draws on chaos theory, fractals and rhizomatic experimentation to explode and disorient movement, De Keersmaeker employs traditional mathematical forms from Euclidean geometry. Her dancers' performance of repeated linear patterns in unison in time to a persistent musical beat is reiterated by the filming of the interactions between De Keersmaeker's choreography and Van der Velde's architecture. Lines of pillars, window and door frames, tiled floors and walls, mirror the lines of dancers and chairs used in the choreography, but the dance fails to be as geometric as at first sight it appears. The linearity of the grid-like patterns is stretched to the hyperbolic limit and beyond when the dancers repeatedly run their hands through their hair and adjust their clothing. Marianne Van Kerkhoven, the company's dramaturg, identified the main theme of *Rosas Danst Rosas* as the contrast and blending of ... 'two voices' – 'the mathematical and the everyday' (1984: 103). De Keersmaeker says that underlying the work is 'the exploration of two seemingly opposite data' (1983). She continues 'these elements, which are not to be separated, are absorbed in a rigorous structure. It is exactly this strictness which shows to full advantage the individual elements' (ibid). These 'individual elements' are themselves caught in grids. The precise repetition of these everyday movements has the effect of cloning, it troubles the notion of an individual discrete subject created by the single viewpoint of perspective and the logic of visualization.

The filming of *Rosas Danst Rosas*, which includes panning, cuts and fast edits, is often disorientating; it interferes with the logic of a single viewpoint which fixes and frames bodies, buildings and space. Cameras move around and through the building filming the dancers between pillars, through glass windows and doorways to spaces beyond. The building and the choreography are perceived on a visual continuum, rather than as a series of fixed images. A sense of distance is created by seeing through spaces, providing multiple viewpoints rather than one that fixes on a single vanishing point. It becomes impossible to map the spaces of the building. When grids of window and door frames are looked through, they and the choreography interfere with and merge with each other, disrupting any sense of perspective. De Keersmaeker reiterates and emphasizes the geometry through the mathematical repetitive grid-like patterns of her choreography. By playing with these grids unfaithfully and stretching their regularity to hyperbolic limits she distracts and disrupts the single viewpoint of perspective and confronts issues of subjectivity.

De Keersmaeker introduces distinctly gendered feminine gestures involving performers touching themselves and occasionally exchanging intimate glances. This familiarity minimizes the distance between them. The precise rhythmic repetition of these gestures, the 'strictness' in De Keersmaeker's terms, absorbs these 'movements and positions borrowed from daily life' (ibid) into the rigorous geometric structure of the choreography, which is in turn absorbed into the spaces of the building. The dancers merge with each other and with the architecture; they become part of the space. They do not appear separate from each other or from the building, they deny spectators a single viewpoint of perspective because their merging with space upsets the logic of visualization which sees subjects represented as separate, individual beings.

Challenging notions of a separated self

Bordo calls the act of seeing from a single viewpoint or perspective 'geometrical seeing', which she argues is dependent on a separation of self from world, subject from object, and in the context of dance, I would add, of spectator from performer/performance. A sense of 'locatedness' in space/time emerges. Bordo argues: 'the development of the human sense of locatedness can be viewed as a process... from which the human being emerges as a... separate entity, no longer continuous with a universe which has... become decisively "other" to it' (1987: 70). This separation of self from world is bound up with Cartesian epistemology, briefly summarized in Chapter 8, which constructs the self as a detached, unified subject. Bordo claims: 'this cold, indifferent universe, and the newly separate self "simply located" within it, form the experiential context for Cartesian epistemological anxiety' which is evident in Descartes' writings as 'a sense of profound distance between self and world' (ibid: 73).

Forsythe and De Keersmaeker minimize and eliminate this distance in their work. Forsythe often includes mime, jazz, musical comedy, sport and theatre in his choreography (Sulcas, 1991a). Hints of street dance are evident in *Enemy in the Figure*. De Keersmaeker's use of non-theatrical spaces, like the RITO school building, minimizes the distance that normally exists between theatrical dance and everyday life. Her *Just Before* (1997) has been performed in Manchester's Nynex Arena, a massive sports stadium, and sets for her pieces often transform regular theatre space making it look non-theatrical. The frame of the stage space in *Enemy in the Figure* is also broken when Forsythe reveals the theatre walls caught in the beam of the moveable lamp.

Forsythe's and De Keersmaeker's dancers' behaviour in performance minimizes distance by breaking conventional theatrical frames. They stand at the side and watch. De Keersmaeker's dancers drink from bottles of water and adjust their hair or clothing. In *Enemy in the Figure* dancers often look as if they are rehearsing or warming up, at one point one throws an item of clothing over the screen. In *Rosas Danst Rosas* the dancers occasionally exchange looks and smile and look directly at the audience. De Keersmaeker said: 'this indicates a relationship . . . to reality' (in Sulcas, 1992: 16).

Van der Velde's practices suggest he had similar views. He designed building interiors, notably Haby's barber shop in Berlin (1901) where water, gas and electric pipes for wiring were exposed (Pevsner, 1960: 101). He criticized the British Arts and Crafts movement interior designer William Morris and his successors for being 'too . . . detached from society' and he claimed that 'rebirth of the arts would emerge from the . . . acceptance of machines and mass production' (Benevolo, 1960: 273). When he directed the School of Applied Art at Weimar, (later the Bauhaus), student products were sold providing 'an unbroken connection with the professional world' (Joedicke, 1961: 43).

Libeskind claimed that 'what . . . Forsythe does with stage space has to do with contemporary sensibility' (1999b). A programme note describes *Enemy in the Figure* as: 'bodies struggling against an environment saturated in technology' (1998) and one reviewer commented that it made her think of 'a world manipulated by the random glare of the media – individuals suddenly exposed in the spotlight, while others . . flail in the shadows' (Mackrell, 1998: 12). Forsythe's use of disorienting deconstructive devices, while interconnecting audience, performers and environments, connotes individuals attuned to the contradictions of their time. As Sulcas noted: 'Forsythe's achievement is to have connected classical dance to the anxious present' (1991a: 7).

In an article subtitled 'Toward a Feminist Architecture', Deborah Fausch (1996) describes architectural projects which she claims as 'feminist' because of the direct bodily experience of physical, historical and/or emotional content that they provide. This emphasis on direct bodily experience she asserts challenges Cartesian dualism that separates self from world, because it foregrounds the importance of physical and sensual perception *other than the visual*, and of bodily connectedness with the world. In architecture, this distinctive feature is identified. In dance, bodily involvement is taken for granted and rarely singled out for discussion. Yet a physical and sensual experience of the world in an obvious sense brings closeness and proximity rather than distance and separation. In the *Rosas Danst Rosas* film, the dancers' physicality, immediately apparent through sounds of their breathing, sighs, and feet scraping and sliding on the tiled floors, creates a proximity between them and their virtual spectators. Fausch also mentions the contribution of architectural features that fail to distance because they are human-sized. One of the noticeable characteristics of Van der Velde's school building is its human-sized proportions. The doorways and window frames are not large and imposing in an institutional sense. These features in *Rosas Danst Rosas* minimize distance by emphasizing corporeality engaging directly with spectators' bodily perception giving them a different, closer experience of space.

Proprioception, which is described as 'sensitivity to stimuli originating in end organ tissues (muscles, tendons, etc.)' (Mattingly, 1999: 27) and 'the awareness of what one feels and sees one's body doing' (Sulcas, 1995a: 8), is a strategy Forsythe employs to engage with spatial and bodily perception by drawing on senses other than the visual. He has likened working in this way to blindness or sight impairment which can intensify proprioception. Dependence on seeing is relinquished in the conventional sense, disorientating the body. This is evident in Forsythe's *Solo* (1995) that he created and danced for the BBC2 series *Evidentia* and which won the 1996 Grand Prix International Video Danse first prize. Forsythe's downward and inward gaze in *Solo* – he sometimes has his eyes closed – suggests that he is drawing inspiration from inner resources rather than from seeing his body in the surrounding space. In this sense he is disoriented and dislocated, or more positively, attuned bodily to the surrounding space in a relaxed yet highly perceptive non-visual manner. There are continual shifts in Forsythe's centre of gravity so that he is often off-balance and falling. His pelvis and hips frequently protrude, forward, behind or to the side, he often goes onto relevé on one foot, only to overbalance and catch himself somewhere else. He makes use of weight, the floor and gravity going down onto one knee and back up, then down onto his back, over onto one hip, then the other and back up. Forsythe's ever shifting weight results in unpredictability and a lack of frontality, his body appears to be continually destabilized. As one critic has stated *Solo* 'explores release from a centrally controlled physicality showing how letting go of something brings [the] unexpected . . . into play' (Nugent 1998: 27). An article about deconstruction on the Ballet Frankfurt website suggests that a deconstructive reading 'encounters and propagates the surprise of otherness' (Johnson, 2000). This surprise of otherness is evident in Forsythe's dislocating and defamiliarizing performance of *Solo*. It is both the effect and the means of deconstruction. Forsythe's dislocated performance disrupts the 'human sense of locatedness' which Bordo suggests usually generates a detached and unified self separate from the world.

Choreographic strategies that challenge the notion of a separated self have been identified in the work of Forsythe and De Keersmaeker. Forsythe's explosion of linearity through disorientation and deconstruction has disrupted the logic of visualization, making it difficult for spectators to remain detached. In order to fathom what is going on they need to engage more directly and bodily with his work. His use of proprioception has introduced an engagement with senses other than the visual, resulting in a radically different encounter of the body with space. De Keersmaeker's reiteration of classical geometrical forms through everyday gestures, viewed against a visual continuum of architecture, not fixed and framed as separate, contributes to the intimacy established in the choreography which minimizes distances between performers and audience. By blurring the boundaries of bodies, buildings and space, these choreographers challenge notions of a separated self, and disrupt the single viewpoint of perspective.

The role of the visual in the construction of bounded and gendered subjectivity

The role of the visual in Western culture is deeply pervasive. It plays a dominant part in Freudian and Lacanian psychoanalytic theories of the construction of the subject where realization of the significance of separation from the (m)other becomes apparent when the small child sees the (m)other as apart and different. This separated self is constructed spatially and visually. According to the French psychoanalyst Jacques Lacan's theory of the 'mirror stage', it is through seeing an image in the mirror that a child experiences itself as a separate being for the first time, what it sees in the mirror is 'a thing with edges' (Minsky, 1996: 144). The image is bounded. In this formative experience space is perceived as distance, as 'we first discover a unified identity from outside ourselves' (ibid: 145).

This visual and spatial construction of subjects as bounded, with edges, is challenged in choreography and architecture that blur the inside/outside boundaries of bodies, buildings and space. The filming of *Rosas Danst Rosas* in the RITO school building, where dancers are viewed through glass windows, foregrounds the transparency of the inside/outside borders of Van der Velde's architecture. The choreography also transgresses these boundaries, taking dancers from inside to outside and vice versa. Libeskind continually plays with these constructions in his architecture. In the Jewish Museum in Berlin he has created empty spaces and placed the moulds of concrete that have been extracted from these voids outside the building. He is literally turning the building inside out. In the museum catalogue he claims, 'building is simultaneously ... external ... and ... internal' (1999a: 34). In *Enemy in the Figure* the wavy lines of the curved wooden screen, reflected in the rope lying on the floor, or when it is undulated by dancers, depart from the linearity of perspective. When the dancers move the rope, and move around and from side to side of both rope and screen, any sense of either of these props becoming a clear boundary or barrier is disrupted. The moveable lamp further emphasizes this by lighting, at different times, both sides of the screen where dancing occurs. There is no sense of front or back, inside or outside, all parts of the stage space can be seen as inside and outside at the same time, confusing these boundaries and challenging the visual and spatial construction of subjects as bounded entities.

In Lacanian psychoanalytic theory the experience of seeing oneself as separate in the mirror also importantly constructs subjects as gendered. When the child sees itself apart from the mother, as having or lacking the phallus, the child becomes gendered. This brings with it a series of implications. According to Lacan, if one has the phallus, with it comes the power and privilege of becoming a human subject, an 'I', but if one lacks the phallus, in Lacan's terms, one is not able to enter fully into subjectivity. Julia Kristeva and Luce Irigaray in their criticism of these theories from a feminist perspective have challenged the dominance of the visual. Kristeva wanted to 'make more detailed the ... stages preceding the mirror stage' because she thought that the grasping of the image by the child involved more than just seeing, drawing on the other senses as well (1984: 22). Irigaray claimed that the visual constitution of the ego in the mirror stage was a blind spot in Lacan's theory that resulted in visual experience being 'inevitably caught in a dialectic of domination in which women were always the victims' (Jay, 1993: 538). Irigaray has stated 'within this logic (... of Western thought), the predominance of the visual ... is particularly foreign to female eroticism. Woman takes pleasure more from touching than from looking, and her entry into a dominant scopic economy signifies ... her consignment to passivity: she is to be the beautiful object of contemplation' (1985b: 26).

Irigaray's emphasis on the role of touching, rather than looking, in female sexuality, providing an alternative to the dominance of the visual, is explored in Chapter 5.

When the dancers in *Rosas Danst Rosas* touch themselves, running hands through hair, or adjusting a tee-shirt, these actions draw attention to the performers' gender and sexuality. These are distinctly feminine gestures; slipping one or both shoulders of a tee-shirt off and back on, throwing a head of long hair forward or backward, cupping a breast with a hand. They are repeated rhythmically and absorbed into the repetitive unison choreographic patterns. Since the dancers are all doing the same, they vividly illustrate the American queer theorist Judith Butler's (1990) claim that gender is a performative act that is learned, rather than constructed through seeing oneself as with or without the phallus. In this sense they challenge the dominance and role of the visual in the construction of gendered subjects. The occasional looks, smiles and nods exchanged between performers, and performers and camera, suggest that they are enjoying themselves and perhaps paradoxically illustrate Irigaray's claim that 'woman takes more pleasure from touching than from looking' (1985b: 26). The dancers' looks to each other and to camera signifying enjoyment operate in different ways. When looking at each other, it is as if they are saying: '"Are you ready? Then here we go"'¹ (Van Kerkhoven, 1984: 103), a sense of camaraderie is expressed. Occasionally looks between performers seem to be more overtly sexual. De Keersmaeker and another dancer perform the same phrase, one standing in the foreground, the other on a raised area behind, they have their backs to each other. They each pull their tee-shirt off a shoulder, pull it back on, turn and look at each other half smiling. They then slip both shoulders of their tee-shirts off and on and exchange glances again as if sharing a sign or code. When dancers look to camera while slipping their tee-shirts off the shoulder or running their hands through their hair, they seem coy or narcissistic, but because they repeat the actions, they are clearly 'performed'. Feminine codes are being played with, resulting in a parody of the kind of femininity constructed by the visual, that Irigaray claims consigns woman to passivity 'to be the beautiful object of contemplation' (1985b: 26).

Forsythe also challenges gender stereotypes in his work. In *Enemy in the Figure* two female dancers perform a duet in unison facing front and kicking their legs doll-like in their automatism, their white socks make them look like American cheer leaders. Like De Keersmaeker, Forsythe seems to be deliberately parodying a female stereotype constructed by the visual. He also employs a variety of different physiques in his company. His female dancers are often very muscular, subverting the 'aesthetic of an androgynously thin, sexless female figure' (Sulcas, 1991a: 33) challenging the scopic regime which consigns women as objects of contemplation. Forsythe has male and female dancers swapping roles and *corps de ballets* composed of both genders performing the same steps underlining 'the corps de ballet's dehumanizing function of endlessly replicating one image (usually that of the heroine)' (ibid: 33). This strategy challenges the dominance of the visual that constructs female subjects as passive and to-be-looked-at.

Ideas about visualization and perspective in philosophy and the role of the visual in the construction of gendered subjects in psychoanalytic theory can be seen to be linked. These links between ocularcentrism and phallocentrism demonstrate the ways in which ideas of perspective and visualization which have privileged a particular way of seeing the world and of experiencing space are bound up with the construction of gendered subjects. When dances and buildings comply with the logic of visualization and act as visual markers or cues which educate people to look in certain ways, they are not just doing this, they are also shaping ways of seeing the world and of constructing space and subjectivity as gendered. De Keersmaeker's and Forsythe's choreography challenges such constructions dependent on the dominance of the visual and a particular way of experiencing space.

Vision and reason – the dominance and construction of the eye/'I'

Links between sight and reason are prevalent in Western thought. 'The mind has been strongly connected with the organ of the eye and the sense of vision. Philosophers have employed vision as a metaphor for thought, and light for the faculty of reason' (Fausch, 1996: 40). 'I see' is synonymous with 'I understand'. The eighteenth-century Enlightenment philosophy – the term 'enlightenment' suggests seeing better – is bound up with rationality; the Enlightenment has been termed the 'age of reason'. Construction of a subject separated from the world, inherited from the Enlightenment, is bound up with the dominance of reason and ideas of representation. The French post-structuralist Jacques Derrida has indicated that the dominance of reason underlies the interpretation of 'beings as objects' 'positioned before a subject' 'as representation', and that the subject who says 'I', ensures his 'mastery over the totality of what is' because representation, protected by reason, dominates (1983: 9–10). The subject thus constructed is not only separate from the world, but certain of himself and his mastery over it and the objects in it. His position is maintained and protected by the dominance of reason, which ensures that this way of seeing the world and subjectivity is accepted. Derrida describes a caricature of this subject, as 'representational man' 'with hard eyes permanently open to a nature that he is to dominate, to rape if necessary, by fixing it' (ibid: 10).

This connection of sight with reason results in the '*logic* of visualization'. This logic arises from and is dependent on the principles of linear perspective. These principles also underpin the classical ballet aesthetic. Classical ballet through its perspectival vocabulary, constructs the body as ideal, separate from the world and uncontaminated by it. Its perfect proportions derive from the lines, angles and forms of Euclidean geometry. This classical body is logically and linearly structured. The logic of the combinations of steps and gestures in the ballet vocabulary, and the rules that connect these, follow geometrical principles based on perspective and vision dependent on unbroken lines. Ballet forms a particular subject according to this logic of clear, visible, geometric occupation of space, and that visual legibility has, following Derrida, social and cultural significance.

Forsythe refuses to present the subject in this way. He deconstructs the classical ballet aesthetic. He considers the unity of the classical ballet body deceptive and illusory. His work is concerned with interrupting 'the mechanics of classical ballet syntax' and with dismantling ballet's assumed, logical structures (Baudoin and Gilpin, 1989: 23). Key to this is his fragmentation of the unbroken lines of the aesthetic. His use of proprioception results in a discontinuous experience of space, a sense of disequilibrium and off-balance, or a shift from ballet's vertical line, evident in tilts of the torso and limbs. Forsythe interrogates the balletic assumptions of balance, placement and verticality. In *Solo* his polycentricity, not only of space but also of the body, results in departure from the linear. He throws his arms behind him one at a time as if trying to get rid of them, they distort his torso, momentarily thrusting his ribcage out. As he tips, turns, twists and gyrates, sometimes jerkily and sometimes fluidly, his moves continually surprise, he looks like he is improvizing. His body never seems coherent or in line, it looks disjointed. Shifts in the centre of gravity result in 'a multicentric agglomerate of points distributed over the body' (Brandstetter, 1998: 48). His dancers' bodies have been described as 'polymorphous figures' (ibid: 47), and his choreographic vocabulary as 'a meandering flow of contortions, which frays in all directions at once and spreads out amoeba-like ... an oscillating construction, fickle and fragile, full of unrest' (Boxberger, 1994:

32). Gilpin, Forsythe's dramaturg in the late eighties, has stated that he 'dismembers the deceptive unity of movement ... in classical ballet, [and] explodes it ... to offer new forms' (1994: 51).

In Forsythe's *Solo* there are no cues for reading the dance linearly, a dance or ballet style is lacking, as is dance or musical phrasing. The choreography is unpredictable, without reference points. Jump cuts and changes of camera angles, together with occasional blackouts or sections of slow motion, exacerbate the unpredictability. These features are typical of Forsythe's work: his *Steptext* (1984) starts before the house lights go down and has sections in complete darkness. At the end of Forsythe's *Solo* his hands are held turned in to the body, a reversal of the classical aesthetic. By challenging the principle of *en dehors* or turn-out in ballet, Forsythe 'plays with this concept of freedom to exploit movement along a multiplicity of planes' (Jackson, 1999: 119). Breaks with the linear logic of the form such as this suggest different ways of experiencing space, which interfere with the logic of visualization that constructs eye/'I' subjects.

De Keersmaeker says of linear logical structures: 'I just think people should stop making sense. If you want to grasp a linear, logical ... understanding of ... events ... on stage, then you get in trouble ... I like it when a performance makes you think of lots of things ... I like the edges. Where you tip over ... I like ... working at the limits' (in Hughes, 1991: 19). Tipping over the edges blurs boundaries of space and subjects and opens up possibilities for experiencing space differently. Linear logical structures result in closed, unidirectional ways of viewing things, but for De Keersmaeker 'closed systems aren't really very credible any more' (ibid: 19). She is more interested in openness, which by working on 'the edges' 'at the limits', defies linear logic and the construction of eye/'I' subjects. Spatial and conceptual openness, central in deconstructive practices, are also key to Forsythe's and Libeskind's work, as Baudoin and Gilpin claim, 'what is paramount to ... Libeskind and Forsythe is, in Libeskind's words, that "as language falls and falters the open is opened"' (1989: 19).

The ideological nature of visualization

Ideas associated with experiencing space through the logic of visualization are not only reductive, as Lefebvre (1991) has indicated, but also ideological. Seeing, more than any other sense, is associated with understanding, with enlightenment and with morality. Bordo points out that perspective has been used as a 'cognitive metaphor' and that 'from its origins in the science of optics' through its development in painting 'connotations of "perspectivity" had been positive' (1987: 123). She notes that *perspicere* means to see clearly and that the Italian poet Alighieri Dante (1265–1321) described perspective as 'the "handmaid" of geometry: "lily-white, unspotted by error and most certain"' (in Baxandall cited in Bordo, 1987: 123). The connotations of whiteness, goodness and certainty have 'racial' and moral implications. Derrida identifies the associations of whiteness and light with visualization, the eye and Western reason. He argues, '[in] the white mythology . . . of the West: the white man takes his own mythology . . . for the universal form of . . . Reason' (1982: 213). This expression of 'racial' domination also derives from privileging 'the sun as the dominant locus of signification' (Jay, 1993: 509). Derrida continues, 'value, gold, the eye, the sun, etc. are carried along . . . by the same tropic movement' (1982: 218). He claims that the sun becomes 'interiorized' in the Westerner who assumes 'the essence of man, "illuminated by the true light"' (ibid: 268). Bordo points out that Peter Limoges 'compares the direct lines of sight of the perspective painting to the clear moral vision of the Godly person' (1987: 123). 'Rectitude', the word for righteousness, correctness or moral uprightness, and 'rectify', which means to put right, come from the same source as 'rectilinear' which means something bounded or characterized by straight lines or forming a straight line. These associations are also importantly gendered. Rectitude is associated with the right side as distinct from the left, and as Robert Hertz suggested in his classic 1909 text *The Pre-eminence of the Right Hand*, 'society and the universe have a whole side which is sacred, noble and precious, and another which is profane and common; a male side, strong and active and another, female, weak and passive; or . . . a right side and a left side' (in Needham, 1973: 10).

Forsythe's use of proprioception, providing a different perception of space, which for him is akin to the experience of sight impairment or blindness, can be seen as an attempt to depart from this ideology of vision bound up with reason, light and rightness associated with the male. His employment of this and other strategies has led to his work being described as an 'architecture of disappearance' by Gilpin, his late eighties' dramaturg. Forsythe makes bodies disappear or become invisible through his use of darkness countering the dominance of light in the ideology of the visual. He also enacts the disappearance of movement by exploring its instability and failure. Failure is key here, Forsythe's polycentricity is a failure to produce linear perspectival work dependent on and creating single viewpoints. His explosion and fragmentation of the body and space result in failure to produce unity. His exploration of bodily instability and disequilibrium result in a failure to maintain the verticality, balance and symmetry of ballet. Gilpin suggests that from these failures come new possibilities for movement, because they describe movement not fixed by meaning, so they 'can be regarded as a positive enabling force of movement' (1994: 50). An article on deconstruction, on Ballett Frankfurt's website in 2000, was entitled 'Nothing fails like success'. It ends

> it is only by forgetting what we know how to do, by setting aside the thoughts that have most changed us, that those thoughts and that knowledge can go on making accessible to us the surprise of an otherness we can only encounter in the moment of suddenly discovering we are ignorant of it.
>
> (Johnson, 2000)

The failures in Forsythe's performances involve the body forgetting how to perform ballet and setting aside that vocabulary in order to make space for the 'surprise of otherness', which is what Gilpin terms a 'positive enabling force of movement'.

These failures create gaps or voids, in-between spaces, moments of disappearance, but these are not negative spaces of absence, they are positive spaces of becoming, in the Deleuzian sense as discussed in Chapter 5. As Gilpin suggests, it is about 'shifting the concept of failure from that of a gap of static space to that of an imperative that causes dynamism' and she likens this to the 'constantly shifting, goalless experience' of becoming (1994: 51). These gaps or voids challenge the dominance of the visual, since in one sense there is nothing there to see. They also challenge the traditional concept or experience of space as empty, because they open up possibilities for perceiving something else that is not fixed but continually in the process of becoming. Forsythe recognizes this when he asserts: 'the whole point of improvisation is to stage disappearance' (in Gilpin, 1994: 52).

Libeskind's architecture works similarly. The voids he has created in the Jewish Museum in Berlin, on one level signify the gaps in Jewish history, on another they mark the disappearance of Jews in the holocaust, but they also suggest positive spaces for present and future becomings. Gilpin claims that Forsythe's choreography similarly focuses on gaps and discontinuities in history, seeing history in a Foucauldian sense. She argues: 'Forsythe focuses on moments of rupture and discontinuity and participates in the sort of history that.... Foucault writes about: "Discontinuity ... has ... become one of the basic elements of historical analysis"'' (ibid: 53). Revealing discontinuities paradoxically involves seeing things in terms of a visual continuum rather than as a series of fixed images. These discontinuities and disappearances give spectators experiences of bodies, space and buildings that are not dependent on the visual, and the ideological associations that accompany it. They require spectators to perceive things differently through direct bodily engagement. The voids in Libeskind's Jewish museum have to be negotiated because they obstruct a straight linear pathway through the building. They create a zig zag across the museum which breaks the line that divided east and west Berlin. By doing this they reveal discontinuities in history and create further discontinuities that move people. As Fausch says of similar architectural projects, which she terms 'feminist' because they require bodily engagement, 'the action of the body must be performed to complete the intellectual content . . . to get the message' and 'this places a higher value upon . . . experience that includes both the body and the mind as opposed to developing one at the expense of the other' (1996: 49). The dominance of the visual, associated with the mind and reason, and with light and rightness, is subverted and replaced by a total body/mind perception of space.

Derrida claimed that associations of visualization with power and the masculine imply violence when he described 'representational man' as 'dominating' or 'raping' nature with 'hard eyes'(1983: 10). The cultural theorist Martin Jay has stated, 'the counterpoint to such violence . . . cannot be another totalizing point of view which would be like a panopticon, but rather fragments, which deny any view of the whole'. He quotes from Derrida's textual accompaniment to a portfolio of photographs by Marie-Françoise Plissart which includes the statement: ' "no single panorama, but simply parts of bodies, torn-up or framed pieces . . . sometimes out of focus, hence blurred" ' (Jay, 1993: 519). This could easily be a description of *Rosas Danst Rosas, Solo* or *Enemy in the Figure*. A 'single panorama' is prevented by the choreography, set, lighting and costume, and in the case of *Rosas Danst Rosas* by the filming also; frequently only parts of bodies are glimpsed. The performance of fragmented movement on multiple planes in all three works eliminates the visibility of the lines of choreography and perspective that lead the eye to see bodies, space and buildings as fixed and framed images. The architect Mark Goulthorpe has indicated that the radical potential of Forsythe's choreography is evident in the way in which he 'breaks up' or diminishes 'the ideology of privileging the eye' (1998).

The moveable lamp in *Enemy in the Figure*, by continually shifting the focus of attention around the stage, emphasizes the polycentricity of

Forsythe's choreography and parodies the importance of the visual, since it operates as a kind of eye/'I' surveying the space. As it turns there are parallels with Foucault's panopticon – the all-seeing viewpoint (see Chapter 7) – but it continually undermines this, since it only sheds light on one part of the stage at a time, leaving the rest unlit. This is partly what Gilpin means when she describes Forsythe's choreography as 'an architecture of disappearance' (1994: 51). The choreography literally disappears from view when it is not lit, or obscured by the set, but its architecture also disappears when the logical structures of the ballet vocabulary, the cues that assist in ordering space and bodies, that determine the way the dance is seen, are no longer apparent. Libeskind has said that Forsythe's work is 'a deepening of an understanding of what the space of the human body really is' (1999b). Gilpin claims that in Forsythe's work 'the absence of vision becomes the culmination of sight' (1994: 53) suggesting that, because this absence disrupts the logic of visualization, new ways of seeing or seeing differently result.

Conclusion

Forsythe and De Keersmaeker have engaged with architectural space differently. Their opposing approaches to geometry – Forsythe's deconstruction and De Keersmaeker's employment of geometric grids – have disrupted the single viewpoint of perspective and the logic of visualization and its associated ideological implications. Their work has suggested new ways of experiencing space and subjectivity, which have been seen as linked through the role of the visual in the construction of spaces and subjects. By recognizing and challenging the links between ocularcentrism and phallocentrism, both choreographers in different ways have suggested possibilities for a subjectivity that no longer has to be gendered and 'racialized', because of its visual and spatial construction as a separated 'I'/eye, dependent on the logic of visualization for its dominance. Forsythe's deconstructed, fragmented and discontinuous bodies and spaces open up possibilities for subjects becoming part of space rather than distanced, contained and unified individuals. De Keersmaeker's dancers, through merging with the architectural spaces of the building and each other, suggest becoming subjects in process, dancer subjects blurring with other dancer subjects. These rethought embodied subjectivities have been revealed through Forsythe's employment of strategies of polycentricity, proprioception, disappearance and failure and De Keersmaeker's use of excessive, precise repetition of a rigorous structure, also occasionally resulting in disappearance. Both foreground direct bodily perception of the world.

The ways in which mutual constructions of space and subjectivity, dancing in in-between spaces and inside/outside bodies and spaces, which have been explored throughout the book, affect subjectivity, have been made apparent in this chapter. In *Enemy in the Figure* and *Rosas Danst Rosas* various reciprocal constructions of space and subjectivity are evident. The space in *Enemy in the Figure* is continually being constructed and reconstructed in specific ways by the dancers, the set, of rope and screen, and the changing lighting. The dancers' fragmented, discontinuous and sometimes explosive experiences of space, and their positioning in relation to set and lighting, also construct them. These mutual constructions open up possibilities for perceiving subjectivity differently through its embodied relationship to space. In *Rosas Danst Rosas* mutual constructions of bodies and spaces in some senses are even more apparent as the geometric patterning of the choreography merges with the geometry of the architecture, but also as the everyday gestures in the choreography connect the dancers to each other. Different kinds of becoming embodied subjects in process are evident in these in-between spaces.

Particular interactions of dancing bodies and certain spaces have been seen throughout the book to create in-between spaces. In *Rosas Danst Rosas* when the dancers are seen filmed through glass windows and walls, in-between spaces are created which it is difficult to map. In this sense in-between spaces disrupt the logic of visualization, challenging the audience's perception and construction of unified spaces and subjects. Spaces in Forsythe's *Enemy in the Figure* are both inside and outside and disappear and reappear because of the combination of set, lighting and choreography. These experiences of in-between spaces in Forsythe's and De Keersmaeker's work suggest new kinds of subjectivity in process. The in-between metaphor allows bodies, spaces and subjects to be conceived and understood as fluid entities that exist in between.

The nature of this fluidity has been explored throughout the book by focusing on body/space interfaces where boundaries are seen to exist. The permeability of these physical boundaries has suggested various blurrings, which challenge the conceptual boundaries evident in binary oppositions and suggest different, interconnected ways of seeing space and subjectivity. Forsythe has blurred the boundaries of bodies and space by exploding linearity through disorientation and deconstruction such that a direct sensual engagement with bodies and space results. De Keersmaeker's dancers, in a very different sense, have been seen to blur with each other and with the architecture of their surroundings.

By examining a range of different postmodern dances through the lenses of various post-structuralist theories, focusing on relationships between dancing bodies and space, it has been possible to show how the dances can challenge entrenched ways of viewing space and subjectivity. Possibilities for rethinking subjectivity 'against the grain' have become evident. Throughout, space has been seen to play a key role in the construction of subjectivity. In this sense, the spatialization of subjectivity has been a focus. Perhaps this focus might not have become evident in quite the same way if dance had not been central. Examining dances, which inevitably involve interactions of actual bodies and spaces, has made fathoming the conceptual complexities involved in rethought notions of space and subjectivity, in some senses, more immediate. The interconnectivity, which has been seen to characterize spatialities of subjectivities, has, on another level, been key to this enterprise. The reciprocal interconnectivity involved in thinking dance through various theories, and thinking theories through various dances, has hopefully facilitated a better understanding of the subject matter of this book – the complex interrelations between dance, space and subjectivity.

Appendix

Dance videos mentioned in the text – notes on availability

Aeroplane Man (1997) choreographed and performed by Jonzi D. Spring Re-Loaded 3 video available from Laban library, Laban, Creekside, London SE8 3DZ. Tel: +44 (0)20 8691 8600 INFO@LABAN.ORG. A 5 min. 2006 version also available on My Space www.myspace.com/jonzid

Between/Outside (1999) filmed and choreographed by Lucille Power, performers: Lucille Power and Sarah Spanton. Available from: Lucille Power, 14c Alvington Crescent, Dalston, London E8 2NW.

Blind Faith (1998) choreographed by Yolande Snaith performed by Yolande Snaith Theatredance. Spring Re-Loaded 4 video record available from Laban library, Laban, Creekside, London SE8 3DZ. Tel: +44 (0)20 8691 8600 INFO@LABAN.ORG

Cross-Channel (1991) choreographed by Lea Anderson, directed by Margaret Williams, MJW production for Channel Four. Available from the BBC and MJW Productions Ltd., 5 Warner House, 43–9 Warner Street, London EC1R 5ER. Tel: 0207 713 0400, Fax: 0207 713 0500.

Duets with Automobiles (1993) made by Terry Braun and Shobana Jeyasingh, performers: Shobana Jeyasingh Dance Company. Commissioned by the Arts Council Film Department and the BBC. Available from Concord Video and Film Distributors, 201 Felixstowe Road, Ipswich, Suffolk, IP3 9BJ. Tel: 01473 726012. www.concordmedia.org.uk

Grotesque Dancer (1987) choreographed by Liz Aggiss and Billy Cowie and performed by Liz Aggiss. Video record available from Concord Video and Film Distributors, 201 Felixstowe Road, Ipswich, Suffolk, IP3 9BJ. Tel: 01473 726012. www.concordmedia.org.uk

Homeward Bound (1997) choreographed and performed by Sarah Spanton. Video record available from Sarah Spanton via www.axisweb.org/artist/sarahspanton

Man (1994) choreographed by Lea Anderson, directed by Margaret Williams. An MJW production for Channel Four. Available from the BBC and MJW Productions Ltd., 5 Warner House, 43–9 Warner Street, London EC1R 5ER. Tel: 0207 713 0400, Fax: 0207 713 0500.

Muurwerk (1987) choreographed and performed by Roxanne Huilmand and directed by Wolfgang Kolb, produced by Éditions à Voir, Amsterdam. Distributed by argos, Werfstraat 13, 1000 Brussels, Belgium. Tel: 32 2 229 00 03. Fax: 32 2 223 73 31 – info@argosarts.org – www.argosarts.org

Outside/In (1995) made by Margaret Williams and Victoria Marks for BBC2 Dance for Camera series commissioned by the Arts Council Film Department and the BBC. Available from Concord Video and Film Distributors, 201 Felixstowe Road, Ipswich, Suffolk, IP3 9BJ. Tel: 01473 726012. www.concordmedia.org.uk

Reservaat (1988) made by Clara Van Gool, performers Martine Berghuijs and Pépé Smit. Distributed by Cinenova Women's Film & Video Distributor, 40 Rosebery Ave., London EC1R 4RX. Email: info@cinenova.org.uk – www.cinenova.org.uk

Rosas Danst Rosas (1997) film by Thierry de Mey, choreography Anne Teresa De Keersmaeker, performers: Rosas Dance Company. Distributed by Total Film Home Entertainment, Dick Lam, PO Box 37743. www.rosas.be

Virginia Minx at Play (1993) choreographed and performed by Emilyn Claid. Available for viewing only from Laban library, Laban, Creekside, London SE8 3DZ. Tel: +44 (0)20 8691 8600 INFO@LABAN.ORG

Notes

1 Introduction

1. With the exception of *Carnets de Traversée*, *Quais Ouest*, *La Deroute* and *Land-Jäger* discussed in Chapter 2, I see all the dances discussed in this book as 'postmodern' in the sense that they recognize the postmodern 'crisis in subjectivity' and present subjects as multiple, embodying differences, fragmented, fluid and open to change. The dances between them also exhibit various other features deemed postmodern such as parody, intertextuality, self-reflexivity, merging of 'high' and 'low' culture and so on.

2 Travel metaphors in dance

1. The power invested in travel discriminates in different ways, as well as gender, 'race', class, sexuality, nationality and other factors all also have a bearing, but the focus here is on gender.
2. Lyotard (1984) for example, sees modernity as characterized by modes of production and postmodernity by modes of information.
3. 'Land-Jäger' can also mean a German sausage, a foot soldier, and quite possibly a traditional dance. It seems likely that the title of the dance video is a play on all of these meanings associated in different ways with dance and masculinity.
4. This style of dancing, consists of recurrent patterns of often violent 'in your face', confrontational gestures and actions, usually performed by young athletic dancers, often clad in Dr Martens boots. It became known through the work of Dutch and Belgian choreographers such as Wim Vandekeybus and Anne Teresa De Keersmaeker (see Chapter 10) in the mid-eighties.
5. I am not using 'mapping' in the sense of Deleuze and Guattari who state: 'what distinguishes the map from the tracing is that it is entirely oriented towards an experimentation in contact with the real. The map does not reproduce an unconscious closed in upon itself; it constructs the unconscious. . . . The map is open and connectable in all of its dimensions . . . [it] has to do with performance, whereas the tracing always involves an alleged "competence"' (1988: 12). My use of 'mapping', which surveys, controls and colonizes, has affinities with their 'tracing'.

Transforming city spaces and subjects

1. Other examples of recent dance texts that use cities as settings or inspirations include *Freefall* (1988) chor.: Gabi Agis, *The Lament of the Empress* (1989) chor.: Pina Bausch (see Chapter 1), *Palermo, Palermo* (1990) chor.: Pina Bausch, *Circumnavigation* (1992) chor.: Norbert and Nicole Corsino, *Topic II* and *49 bis* (1992) chor.: Sarah Denizot, *Duets with Automobiles* (1993) chor.: Shobana

Jeyasingh (see Chapter 6) and *Dark Hours and Finer Moments* (1994) chor.: Gabi Agis. For discussion of *Freefall, The Lament of the Empress, Circumnavigation, Topic II*, and *Dark Hours and Finer Moments* see Briginshaw (1997).
2. *Street Dance* was performed again in 1965 at Robert Rauschenberg's studio and the transcript of the accompanying tape was included in 'Lucinda Childs: A Portfolio' in *Artforum* 11 February 1973 and reprinted in Banes, 1980: 146–7. The transcript gives precise details of signs, lettering and contents of window displays that the dancers referred to in their performance.
3. As McDonagh's interpretation of *Blueprint* indicates (1990: 115).
4. Space traditionally tends to be feminized deriving from Plato's notions of the female *chora*. In Plato's view space is conceptualized as 'a bounded entity' 'a sort of container' associated with the female body particularly with that of the mother (Best, 1995: 182). He argues 'it [the receptacle/space] . . . is a kind of . . . plastic material on which changing impressions are stamped by the things which enter it making it appear different at different times . . . we may use the metaphor of birth and compare the receptacle to the mother' (quoted in Best, 1995, p.184).

4 Coastal constructions

1. Gisbourne's comment forms part of his discussion of the British visual artist Marc Quinn's mercury-like blown glass sculptures *Morphologies* (1996–98), which have been described as mutations. There are certain parallels with the mutating couples in *Out on the Windy Beach*.
2. Anderson also plays with mutations in her *Yippeee!!!* (2006) which we discuss in some detail in Briginshaw and Burt (2009).

5 Desire spatialized differently

1. Clara Van Gool also directed the television version of the British physical theatre company DV8's *Enter Achilles* (1995) choreographed by Lloyd Newson.
2. Recent festival screenings of *Reservaat* include *On the bend Film Festival* (Canada), June 1998; *Outfest*, Los Angeles, July 1998; *Out on Screen*, Vancouver, August 1998; *The London Lesbian and Gay Film* Festival, April/May 1999; *Inside Out*, Toronto, May 1999; *Making Scenes Film and Video Festival*, Ottawa September 1999.
3. It is important to distinguish between Butler's use of repetition seen as 'unfaithful' and Deleuze's use of repetition which is slightly more positive because he sees repetition as never the same, always producing difference rather than lack. However, because Butler has usefully developed the concept of performativity of gender and sexuality, which Deleuze has not, in this sense her work is pertinent for my purposes. For more discussion and exploration of the values of unfaithful repetition of performative acts see our discussion of Lea Anderson's *Yippeee!!!* in Briginshaw and Burt (2009).

6 Hybridity and nomadic subjectivity

1. When discussing *Duets with Automobiles*, Jeyasingh stated she would no longer use the word 'icon', that it was too closely connected with a particular traditional image of an Indian dancer. She has 'moved on' from that idea. She also

said that if she were making *Duets with Automobiles* today she would change the costume and make-up; removing the *bhindis* from the dancers' foreheads. The costume she would make less silky and in a less bright colour. She stated that it would be enough for her that the dancers looked Indian through their skin colour (Jeyasingh, 1997a).
2. In interview, Jeyasingh wanted to emphasize the important role played by Braun, the director and filmmaker, in the creation of *Duets with Automobiles*, which she considered was a true collaboration. She stated: 'Terry gave me the room to make... what I wanted, he also wanted the same thing' (Jeyasingh, 1997a). The film was made totally *in situ* and in dialogue with the camera and Jeyasingh was present in the editing room throughout. To indicate this collaboration, Braun and Jeyasingh share the same frame in the title credits.
3. Terry Braun is the filmmaker for both *Step in Time Girls* and *Duets with Automobiles* and certain similarities in style are apparent.

7 Crossing the (black) Atlantic

1. For an examination of the powerful role of tableau in art and dance, see Chapter 8.
2. Bob Rosen was producer, co-director and editor of the film.
3. 'Breaking' or 'break dance' are terms coined by the media around 1984 when breaking became popular, but the performers themselves prefer to be called B Boys. The 'B' refers to the break in the rhythm section of the music which they respond to in the dance (Ogg and Upshal, 1999).

8 Fleshy corporealities

1. Lea Anderson trained at St Martins School of Art and Yolande Snaith at Wimbledon School of Art.
2. The duet is variously titled *You can see us* and *You can see me*. There is some video footage of rehearsals for this with Bill T. Jones.
3. *Blind Faith* had two casts due to Snaith's pregnancy during the latter half of 1998. The first performance with the first cast was at the Nuffield Theatre, Lancaster on 3 February 1998. After the first tour, the piece was rehearsed with a second cast. Henrietta Hale took Snaith's part and Ruth Spencer took Clayden's role, and a duet between Snaith and Clayden was transferred to Hale and Russell Trigg. The rest of the cast remained the same and the dance was substantially the same. The second cast gave its first performance in Germany on 12 October 1998.
4. Here 'mapping' is used in the Deleuzian sense – see Chapter 2 note 5, and not in the sense in which it is referred to in Chapter 2.
5. The term 'jouissance' has no equivalent in English, the nearest translation is the bliss of sexual orgasm. Kristeva's use of the term refers to a specifically feminine form of excess (see Wright, 1992: 185–8).
6. *Chora* derives from Plato and is a general term for 'place', 'site' or 'receptacle', associated with the maternal functions of femininity. For Kristeva, it is 'the site of undifferentiated bodily space' for 'the production of the ... womb and matter' that mother and child share (Grosz in Wright, 1992: 195).

9 'Carnivalesque' subversions

1. Except the binary opposition of carnival/normal social life on which these oppositions depend.
2. *Sprechgesang* – a theatrical cabaret genre of spoken song often found in Kurt Weill/Bertold Brecht musical theatre and music hall, for example.
3. Although there are examples of the blurring of high and low culture before postmodernism, the context is different. Postmodern works deliberately blur these categories in order to emphasize their refusal to operate traditional moral–aesthetic valuations (Wheale, 1995: 35).
4. Bakhtin uses the word 'gay' throughout his text to refer to the joyfulness of festivals and carnivals, and to the 'fullness', 'mirth' or 'immorality' of much festival, marketplace and carnival behaviour. Bakhtin is not using 'gay' to mean homosexual – that use has come about more recently, initially as a slang term or euphemism, and then reclaimed as a positive descriptor. Bakhtin's use of 'gay' meaning fullness or overindulgence, often related to sexual promiscuity, is a telling reminder of the word's origins. It also demonstrates the openness of language that Bakhtin theorizes.
5. See note 4.

Bibliography

Acocella, J. *Mark Morris* New York: Farrar Strauss Giroux, 1993.
Adair, C. 'Seminars at the 10th Dartington Dance Festival' *New Dance* No. 41 Summer 1987, 20.
Aditi Newsletter 'Critical Debate around Bharata Natyam in Britain: Old Paths, New Directions' *Aditi Newsletter*, Spring 1997, 8–9.
Ainley, R. (ed.) *New Frontiers of Space, Bodies and Gender* London: Routledge, 1998.
Albright, A.C. *Choreographing Difference* Hanover: Wesleyan University Press, 1997.
Anderson, L. Interview with the author, June 1997.
Anderson, L. Personal communication with the author, February 2000.
Aronson, A. *The History and Theory of Environmental Scenography* Ann Arbor, Michigan: UMI Research Press, 1981.
Ashcroft, B., Griffiths, G. and Tiffin, H. (eds) *The Post-Colonial Studies Reader* London: Routledge, 1995.
Baker, R. 'New Worlds for Old: The Visionary Art of Meredith Monk' *American Theatre* Vol. 1 No. 6, 1984, 1–3.
Bakhtin, M.M. *The Dialogic Imagination* trans. C. Emerson and M. Holquist, Austin: University of Texas, 1981.
Bakhtin, M.M. *Rabelais and his World* trans. H. Iswolsky, Bloomington: Indiana University Press, 1984.
Banes, S. *Terpsichore in Sneakers* Boston: Houghton and Mifflin, 1980.
Banes, S. *Greenwich Village 1963* Durham and London: Duke University Press, 1993.
Barker, F. *The Tremulous Private Body* Ann Arbor: University of Michigan Press, 1995.
Barthes, R. *Image Music Text* trans. S. Heath, London: Fontana, 1977.
Battersby, C. 'Her Body/Her Boundaries Gender and the Metaphysics of Containment', *Journal of Philosophy and the Visual Arts*, Special Issue on the Body, 1993, 30–9.
Baudoin, P. and Gilpin, H. 'Proliferation and Perfect Disorder: William Forsythe and the Architecture of Disappearance' *Parallax* Frankfurt, 1989, 9–25.
Bell, D. and Valentine, G. (eds) *Mapping Desire*, London: Routledge, 1995.
Benevolo, L. *History of Modern Architecture* London: Routledge and Kegan, Paul, 1960.
Berger, J. *Ways of Seeing* London: BBC Books, 1973.
Best, S. 'Sexualising Space', in Grosz, E. and Probyn, E. (eds) *Sexy Bodies*, London: Routledge, 1995.
Betterton, R. *An Intimate Distance Women, Artists and the Body* London: Routledge, 1996.
Bhabha, H. 'The Other Question: Difference, Discrimination and the Discourse of Colonialism', in Ferguson, R., Gever, M. Minh-ha, Trinh T. and West, C. (eds) *Out There: Marginalization and Contemporary Culture* Cambridge, Mass.: MIT Press, 1990.

Bhabha, H.K. 'The Third Space', in Rutherford, J. (ed.) *Identity, Community, Culture, Difference* London: Lawrence and Wishart, 1991.
Bhabha, H.K. *The Location of Culture* London: Routledge, 1994.
Bhabha, H.K. 'Culture's in-between', in Bennett, D. (ed.) *Multicultural States. Rethinking Difference and Identity* London: Routledge, 1998.
Bird, J., Curtis, B., Puttnam, T., Robertson, G. and Tickner, L. (eds) *Mapping the Futures* London: Routledge, 1993.
Birkett, D. 'Have guilt, will travel', *New Statesman and Society* 13 June 1990, 41–2.
Blind Faith Programme, 1998.
Bordo, S.R. *The Flight to Objectivity* Albany N.Y.: State University of New York, 1987.
Bordo, S. 'Feminism, Foucault and the Politics of the Body', in Ramazanoglu, C. (ed.) *Up Against Foucault* London: Routledge, 1993.
Bordo, S. *Unbearable Weight Feminism, Western Culture and the Body* Berkeley, CA: University of California Press, 1995.
Boxberger, E. 'Want To be Hypnotized' *Ballett International/Tanz Aktuel* February 1994, 28–32.
Boxberger, E. 'The Body is Not Only Objectivity', *Ballett International/Tanz Aktuel*, February 1997, 24–5.
Brah, A. *Cartographies of Diaspora* London: Routledge, 1996.
Braidotti, R. *Nomadic Subjects* New York: Columbia University Press, 1994.
Brandstetter, G. 'Defigurative Choreography from Marcel Duchamp to William Forsythe' *The Drama Review* 42:4 (T160) Winter 1998, 37–55.
Briginshaw, V.A. 'Do we really know what post-modern dance is?' *Dance Theatre Journal* 6:2 Spring 1988, 12–13, 24.
Briginshaw, V.A. 'The Wiggle Goes On – a profile of Liz Aggiss and Billy Cowie' *New Dance* No. 44 June 1988, 7–11.
Briginshaw, V.A. 'Postmodern dance and the politics of resistance', in Campbell, P. (ed.) *Analysing Performance A Critical Reader* Manchester: Manchester University Press, 1996.
Briginshaw, V.A. '"Keep Your Great City Paris!" – the Lament of the Empress and Other Women', in Thomas, H. (ed.) *Dance in the City*, London: Macmillan – now Palgrave, 1997.
Briginshaw, V.A. 'Theorising the Performativity of Lesbian Dance', *Society for Dance History Scholars Conference Proceedings*, University of Oregon, 1998.
Briginshaw, V.A. and Burt, R. *Writing Dancing Together*, Basingstoke: Palgrave, 2009.
Brown, T. unpublished talk given at the *Northern Electric Light* Festival, Newcastle-upon-Tyne, 1996.
Burch, N. *Theory of Film Practice* Princeton, 1968.
Burgin, V. 'Diderot, Barthes, *Vertigo*', in Burgin, V., Donald, J. and Kaplan, C. (eds) *Formations of Fantasy* London: Methuen, 1986.
Burgin, V. *In/Different Spaces Place and Memory in Visual Culture* Berkeley CA: University of California Press, 1996.
Butler, J. *Gender Trouble* New York: Routledge, 1990.
Butler, J. 'Performative Acts and Gender Constitution', in Case, S.-E. (ed.) *Performing Feminisms* Baltimore: Johns Hopkins University Press, 1990.
Butler, J. *Bodies that Matter* London: Routledge, 1993.
Butler, J. *The Psychic Life of Power* Stanford, CA: Stanford University Press, 1997
Carlson, M. *Places of Performance. The Semiotics of Theatre Architecture* Ithaca: Cornell University Press, 1989.

Carney, R. *Speaking the Language of Desire. The Films of Carl Dreyer*, Cambridge University Press, New York, 1989.
Chambers, I. 'Signs of Silence, Lines of Listening', in Chambers, I. and Curti, L. (eds) *The Post-Colonial Question* London: Routledge, 1996.
Chaney, D. *The Cultural Turn* London: Routledge, 1994.
Chaudhuri, U. *Staging Place. The Geography of Modern Drama* Ann Arbor, Michigan: University of Michigan Press, 1995.
Clarke, D.B. *The Cinematic City*, London: Routledge, 1997.
Clifford, J. 'Traveling cultures', in Grossberg, L., Nelson, C. and Treichler, P. (eds) *Cultural Studies*, London: Routledge, 1992.
Cohen, S.J. *Dance as a Theatre Art* Princeton, NJ: Dance Horizons, 1992.
Colomina, B. (ed.) *Sexuality and Space* Princeton, NJ: Princeton University Press, 1992.
Connor, S. *Postmodernist Culture* Oxford: Basil Blackwell, 1989.
Constanti, S. 'Easing the Load The Spring Loaded Season at the Place' *Dance Theatre Journal* 5:2 Summer 1987: 26.
Crow, T. *The Rise of the Sixties* London: Weidenfeld and Nicholson, 1996.
De Certeau, M. *The Practice of Everyday Life* trans. S. Rendall, Berkeley CA: University of California Press, 1988.
De Keersmaeker, A.T. 'Valeska Gert' *The Drama Review* 25(3) Fall 1981, 55–67.
De Keersmaeker, A.T. *Rosas Danst Rosas Presentation* Rosas Danst Rosas, 1983. www.rosas.be
Deleuze, G. *Cinema 1 The Movement Image*, trans. H. Tomlinson and B. Habberjam, London: The Athlone Press, 1992.
Deleuze, G. *The Fold Leibniz and the Baroque*, trans. T. Conley, London: The Athlone Press, 1993.
Deleuze, G. *Negotiations* trans. M. Joughin New York: Columbia University Press, 1995.
Deleuze, G. and Guattari, F. *Kafka: Toward a Minor Literature* trans. D. Polan, Minneapolis: Minneapolis University Press, 1986.
Deleuze, G. and Guattari, F. *A Thousand Plateaus Capitalism and Schizophrenia* trans. B. Massumi, London: The Athlone Press, 1988.
Deleuze, G. and Guattari, F. *What is Philosophy?* trans. H. Tomlinson and G. Burchell, London: Verso, 1994.
Derrida, J. *Margins of Philosophy*, trans. A. Bass, Chicago: Chicago University Press, 1982.
Derrida, J. 'The Principle of Reason: The University in the Eyes of its Pupils', *Diacritics* Fall 1983, 3–21.
Devi, M. *Imaginary Maps* trans. G. Spivak, New York: Routledge, 1995.
Diprose, R. and Ferrell, R. (eds) *Cartographies Postructuralism and the Mapping of Bodies and Spaces* Australia: Allen and Unwin, 1991.
Docherty, T. (ed.) *Postmodernism A Reader* London: Harvester Wheatsheaf, 1993.
Driver, S. '2 or 3 Things That Might Be Considered Primary', *Ballet Review* 18:1, 1990a, 81–5.
Driver, S. and the editors of *Ballet Review* 'A Conversation with William Forsythe', *Ballet Review* 18:1, 1990b, 86–97.
Duncan, N. (ed.) *Body Space* London: Routledge, 1996.
Eagleton, T. *Walter Benjamin: Towards a Revolutionary Criticism* London: Verso, 1981.
Éditions à Voir, Video publicity material for *Land-Jäger* (1990) Amsterdam: Éditions à Voir, 1990.

Éditions à Voir, Video cover publicity material for *Carnets de Traversée, Quais Ouest* (1989, video: 1990), *La Deroute* (1990, video: 1991), Amsterdam: Éditions à Voir, 1991.
Ellis Island publicity flier *The Award Winning Film 'Ellis Island'*, New York: Greenwich Film Associates, 1981.
Farman, V. 'Grotesque Dancer Liz Aggiss, Dartington Dance Festival' *New Dance* No. 41 Summer 1987, 20.
Fausch, D. 'The Knowledge of the Body and the Presence of History – Towards a Feminist Architecture', in Coleman, D., Danze, E. and Henderson, C. (eds) *Architecture and Feminism* 9 New York: Princeton Architectural Press, 1996.
Feliciano, R. 'Trisha Brown', *Dance Now*, 5:1, 1996, 7–11.
Feliciano, R. 'Anne Teresa De Keersmaeker: a love-hate affair with dance', *Dance Magazine* LXXII No. 3 March 1998, 96.
Florence, P. 'Innovation, sex and spatial structure', in Lagerroth, U., Lund, H. and Hedling, E. (eds) *Interart Poetics. Essays on the Interrelations of Arts and Media*, Amsterdam: Editions Rodopi, 1998.
Forsythe Study Day Lectures, Sadler's Wells Theatre, London, 28 November, 1998.
Foucault, M. *Discipline and Punish* trans. A. Sheridan. Harmondsworth: Peregrine, 1977.
Foucault, M. 'The Subject and Power', in Dreyfus, H. and Rabinow, P. (eds) *Michel Foucault: Beyond Structuralism and Hermeneutics* Chicago: University of Chicago Press, 1982.
Foucault, M. Of 'Other Spaces' *Diacritics* 16, No. 1 Spring, 1986.
Gilbert, H. 'Dance, Movement and Resistance Politics', in Ashcroft, B., Griffiths, G. and Tiffin, H. (eds) *The Post-Colonial Studies Reader* London: Routledge, 1995.
Gilbert, J. 'Forsythe's generation is game' *The Independent on Sunday* 29 November 1998, 13.
Gilpin, H. 'Aberrations of Gravity', *Any* (Architecture New York) 5: Lightness, March/April 1994, 50–5.
Gilroy, P. *The Black Atlantic* London: Verso, 1993.
Gilroy, P. '"... to be real": the dissident forms of black expressive culture', in Ugwu, C. (ed.) *Let's Get It On: The Politics of Black Performance* London: ICA, Seattle: Bay Press, 1995.
Gilroy, P. 'Route Work: The Black Atlantic and the Politics of Exile', in Chambers, I. and Curti, L. (eds) *The Post-Colonial Question* London: Routledge, 1996.
Gilroy, P. 'Diaspora, Utopia and the Critique of Capitalism', in Gelder, K. and Thornton, S. (eds) *The Subcultures Reader* London: Routledge, 1997.
Ginot, I., 'The Power of Inner Poetics', *Ballett International/Tanz Aktuel*, February, 1997, 20–3.
Gisbourne, M. '*Dis*-incarnate', in Quinn, M. *Marc Quinn Incarnate* London: Booth-Clibborn Editions, 1998.
Gottschild, B.D. *Digging the Africanist Presence in American Performance* Westport, CT: Praeger, 1996.
Goulthorpe, M. 'An Architecture of Disappearance' unpublished paper given at the *Forsythe Study Day Lectures*, Sadler's Wells Theatre, London, 28 November 1998.
Griggers, C. 'Lesbian Bodies in the Age of (Post)Mechanical Reproduction', in Doan, L. (ed.) *The Lesbian Postmodern* New York: Columbia University Press 1994.

Gross, E. 'The Body of Signification', in Fletcher, J. and Benjamin, A. (eds), *Abjection, Melancholia and Love*, London: Routledge, 1990.
Grossberg, L. 'The Space of Culture, The Power of Space', in Chambers, I. and Curti, L. (eds) *The Post-Colonial Question*, London: Routledge, 1996.
Grosz, E. *Sexual Subversions Three French Feminists* Australia: Allen and Unwin, 1989.
Grosz, E. 'Judaism and Exile: The Ethics of Otherness', *New Formations*, 12 1990, 77–88.
Grosz, E. 'Refiguring Lesbian Desire', in Doan, L. (ed.) *The Lesbian Postmodern* New York: Columbia University Press, 1994a.
Grosz, E. *Volatile Bodies* Bloomington, Indiana: Indiana University Press, 1994b.
Grosz, E. *Space, Time and Perversion* London: Routledge, 1995.
Haraway, D. 'A Manifesto for Cyborgs', in Nicholson, L. (ed.) *Feminism/Postmodernism*, New York: Routledge, 1990.
Haraway, D. *Simians, Cyborgs and Women* London: Free Association Books, 1991.
Haraway, D. 'The Promises of Monsters: a regenerative politics for inappropriate/d others', in Grossberg, L., Nelson, C. and Treichler, P. (eds) *Cultural Studies* London: Routledge, 1992.
Haraway, D. *Modest_Witness@Second_Millenium.Female Man©_Meets_OncoMouse^{TM}* New York: Routledge, 1997.
Harris, R. 'Notes from Rennie Harris of Pure Movement' *'Festival of Four Worlds'* program, Philadelphia, Spring, 1994.
Harrison, C. and Wood, P. (eds) *Art in Theory 1900–1990*, Oxford: Blackwell, 1992.
Harvey, D. *The Condition of Postmodernity* Oxford: Blackwell, 1990.
Heelan, P. *Space-Perception and the Philosophy of Science*, Berkeley CA: University of California Press, 1983.
Hill, L. 'Julia Kristeva Theorizing the Avant-Garde?', in Fletcher, J. and Benjamin, A. (eds) *Abjection, Melancholia and Love, the work of Julia Kristeva* London: Routledge, 1990.
Holmstrom, L. 'Introduction. In Pursuit of the New', in *'romance...with footnotes' monograph*. London: Shobana Jeyasingh Dance Company, 1995.
hooks, b. 'Marginality as Site of Resistance', in Ferguson, R., Gever, M.Minh-ha, Trinh T. and West, C. (eds) *Out There: Marginalization and Contemporary Culture*, Cambridge, Mass.: MIT Press, 1990.
Howard, R. 'CandoCo' *Dancing Times* January 1998, 315.
Hughes, D. 'Stop making Sense. David Hughes talks to Anne Teresa De Keersmaeker about "Stella"' *Dance Theatre Journal* 9:1 Summer 1991, 16–19.
Huyssen, A. 'Mapping the Postmodern' *New German Critique* No. 33 Fall 1984, 5–52.
Ingram, L. 'An Interview with Shobana Jeyasingh', *British Studies Now* Issue 9, April 1997, 10–12.
Irigaray, L. 'Interview', in Hans, M.F. and Lapouge, G. (eds) *Les Femmes, La Pornographie, L'Erotisme* Paris: Seuil, 1978.
Irigaray, L. *Speculum of the Other Woman* trans. G.C. Gill, London: Athlone Press, 1985a.

Irigaray, L. *This Sex Which is Not One* trans. C. Porter with C. Burke, London: Athlone Press, 1985b.
Irigaray, L. *An Ethics of Sexual Difference* trans. C. Burke and G.C. Gill, London: Athlone Press, 1993.
Jackson, J. 'William Forsythe's Challenge to the Balletic Text: Dancing Latin', in Adshead-Lansdale, J. (ed.) *Dancing Texts*, London: Dance Books, 1999.
Jameson, F. 'Cognitive Mapping', in Nelson, C. and Grossberg, L. (eds) *Marxism and the Interpretation of Culture* Urbana and Chicago: University of Illinois Press, 1988.
Jameson, F. *Postmodernism or the Cultural Logic of Late Capitalism* London, Verso, 1991.
Jarvis, B. *Postmodern Cartographies*, London: Pluto, 1998.
Jay, M. *Downcast Eyes. The Denigration of Vision in Twentieth Century French Thought*, Berkeley: University of California Press, 1993.
Jelavich, P. *Berlin Cabaret* Cambridge, MA: Harvard University Press, 1993.
Jeyasingh, S. Shobana Jeyasingh speaking on the *Making of Maps* video, London: Shobana Jeyasingh Dance Company, 1993.
Jeyasingh, S. Shobana Jeyasingh speaking on the *Romance . . . with footnotes* video, London: Shobana Jeyasingh Dance Company, 1994.
Jeyasingh, S. 'Imaginary Homelands: Creating a New Dance Language', in Jones, C. and Lansdale, J. (eds) *Border Tensions: Dance and Discourse* Guildford, Surrey: University of Surrey, 1995.
Jeyasingh, S. *So Many Islands Seminar*, The Place, London, 1 November, 1996.
Jeyasingh, S. Shobana Jeyasingh in interview with the author, May 1997a.
Jeyasingh, S. 'Text Context Dance', *Choreography and Dance* Vol. 4(2) 1997b, 31–4.
Joedicke, J. *A History of Modern Architecture* London: Architectural Press, 1961.
Johnson, B. 'Nothing fails like success' downloaded from *Ballet Frankfurt* 2000. www.frankfurtballett.de
Jonzi D. Interview on *Aeroplane Man* video. The Place Spring Loaded 3 series, 1997.
Kael, P. 'Review of *The Passion of Joan of Arc*' in *Kiss Kiss Bang Bang* London: Calder and Boyars, 1970.
Kasbarian, J.A. 'Mapping Edward Said: geography, identity and the politics of location', in *Environment and Planning D: Society and Space* 14(4) 1996, 529–58.
Kent, S. 'Feminism and Decadence', *Artscribe* No. 47, 1984, 54–61.
Kreemer, C. *Further Steps* N.Y.: Harper and Row, 1987.
Kristeva, J. *Powers of Horror. An Essay on Abjection* trans. L.S. Roudiez, N.Y.: Columbia University Press, 1982.
Kristeva, J. 'Julia Kristeva in Conversation with Rosalind Coward', in *ICA Documents*, special issue on *Desire* London: ICA, 1984.
Kristeva, J. *Tales of Love* trans. L.S. Roudiez, New York: Columbia University Press, 1987.
Kristeva, J. *Black Sun* trans. L.S. Roudiez, New York: Columbia University Press, 1989.
Leask, J. 'Downtown London: the new artists' *Ballett International/Tanz Aktuel* 6/97 1997, 36–9.

Leask, J. 'Jonzi D' *Dance Theatre Journal* 14:1 1998, 46–7.
Lee, R. 'Resisting Amnesia: Feminism, Painting and Postmodernism' *Feminist Review* No. 26 Summer 1987, 5–28.
Lefebvre, H. *The Production of Space* trans. D. Nicholson-Smith, Oxford: Blackwell, 1991.
Lefebvre, H. *Writings on Cities* trans. and edited E. Kofman and E. Lebas, Oxford: Blackwell, 1996.
Lepecki, A. 'The Liberation of Space', *Ballett International/TanzAktuel*, February, 1997, 14–19.
Libeskind, D. *Jewish Museum Berlin* Berlin: G and B Arts International 1999a.
Libeskind, D. in *William Forsythe. I'm seeing your finger as a line* interviewed by Christopher Cook for BBC Radio 3 broadcast 14 March, 1999b.
Lippard, L. *The Lure of the Local* New York: New Press, 1997.
Livet, A. (ed.) *Contemporary Dance* New York: Abbeville Press, 1978.
Lorraine, T. *Irigaray and Deleuze Experiments in Visceral Philosophy* Ithaca and London: Cornell University Press, 1999.
Lyotard, J. *The Postmodern Condition* trans. G. Bennington and B. Massumi, Manchester: Manchester University Press, 1984.
Mackrell, J. 'Ballett Frankfurt' *The Guardian* 26 November, 1998, 12.
Mallems, A. 'The Belgian Dance Explosion of the Eighties' *Ballett International* 2/91 1991, 19–24.
Massey, D. 'Flexible Sexism' *Environment and Planning D: Society and Space* Vol. 9 1991, 31–57.
Massey, D. 'Power-geometry and a progressive sense of place', in Bird, J., Curtis, B., Puttnam, T., Robertson, G. and Tickner, L. (eds) *Mapping the Futures* London: Routledge, 1993.
Massey, D. *Space, Place and Gender* Cambridge: Polity Press, 1994.
Mattingly, K. 'Deconstructivists Frank Gehry and William Forsythe: De-Signs of the Times' *Dance Research Journal* 31/1 Spring 1999, 20–8.
Mazo, J. *Prime Movers* London: A. and C. Black, 1977.
McDonagh, D. *The Rise and Fall and Rise of Modern Dance* NJ. Pennington: a cappella books, 1990.
McNay, L. *Foucault and Feminism*, Cambridge: Polity Press, 1992.
Meisner, N. 'Same difference', *Dance Now* 8:3 1999, 70–3.
Merleau-Ponty, M. *The Phenomenology of Perception*, trans. C. Smith, London: Routledge and Kegan Paul, 1962.
Minsky, R. (ed.) *Psychoanalysis and Gender* London: Routledge, 1996.
Morris, M. 'Future Fear', in Bird, J., Curtis, B., Puttnam, T., Robertson, G. and Tickner, L. (eds) *Mapping the Futures* London: Routledge, 1993.
Morris, P. (ed.) *The Bakhtin Reader* London: Edward Arnold, 1994.
Mulvey, L. 'Visual Pleasure and Narrative Cinema' *Screen* Vol. 16, No. 3, 1975, 6–18.
Mulvey, L. *Visual and Other Pleasures* London: Macmillan (now Palgrave), 1989.
Nast, H. and Pile, S. *Places through the Body* London: Routledge, 1998.
Needham, R. (ed.) *Introduction to Right and Left: Essays on Dual Symbolic Classification* Chicago: University of Chicago Press, 1973.
Norberg-Schulz, C. *Existence, Space and Architecture* London: Studio Vista, 1971.

Norberg-Schulz, C. *Genius Loci. Towards a Phenomenology of Architecture* New York: Rizzoli International Publications, 1980.
Nugent, A. 'Eyeing Forsythe' *Dance Theatre Journal* 14:3 1998, 26–30.
Odenthal, J. 'A Conversation with William Forsythe' *Ballett International/Tanz Aktuel* 2/94 1994, 33–7.
Ogg, A. and Upshal, D. *The Hip Hop Years. The History of Rap.* London: Channel Four Books, 1999.
Parker, A 'Writing against Writing and Other Disruptions in Recent French Lesbian Texts', in Kauffman, L. (ed.) *Feminism and Institutions* Oxford: Blackwell, 1989.
Patton, P. 'Nietzsche and the body of the philosopher', in Diprose, R. and Ferrell, R. (eds) *Cartographies Poststructuralism and the Mapping of Bodies and Spaces* Australia: Allen and Unwin, 1991.
Patton, P. 'Imaginary Cities: Images of Postmodernity', in Watson, S. and Gibson, K. (eds) *Postmodern Cities and Spaces* Oxford: Blackwell, 1995.
Paxton, S. 'Letter to Trisha' *Contact Quarterly*, Winter/Spring, 1995, 94.
Perron, W. 'Trisha Brown on Tour', *Dancing Times*, May 1996, 749–51.
Pevsner, N. *Pioneers of Modern Design From William Morris to Walter Gropius* London: Penguin Books, 1960.
Pile, S. *The Body and the City* London: Routledge, 1996.
Pollock, G. *Generations and Geographies in the Visual Arts* London: Routledge 1996.
Power, L. unpublished interview with the author, February 1999.
Probyn, E. 'Travels in the postmodern: making sense of the local', in Nicholson, L.J. (ed.) *Feminism/Postmodernism* New York: Routledge, 1990.
Rabinow, P. (ed.) *The Foucault Reader* London: Penguin, 1984.
Read, A. *Theatre and Everyday Life* London: Routledge, 1993.
Rose, G. 'Some notes towards thinking about the spaces of the future', in Bird, J., Curtis, B., Puttnam, T., Robertson, G. and Tickner, L. (eds) *Mapping the Futures* Routledge, London, 1993a.
Rose, G. *Feminism and Geography* Cambridge: Polity Press, 1993b.
Roy, S. 'Dirt, Noise, Traffic: Contemporary Indian Dance in the Western City; Modernity, Ethnicity and Hybridity', in Thomas, H. (ed.) *Dance in the City*, London: Macmillan (now Palgrave), 1997.
Rubidge, S. *Shobana Jeyasingh Dance Company 'Making of Maps' Resource Pack*, London: Shobana Jeyasingh Dance Company, 1993.
Rubidge, S. *romance . . . with footnotes monograph*, London: Shobana Jeyasingh Dance Company, 1995.
Rushdie, S. *Imaginary Homelands: Essays and Criticism 1981–1991* London: Granta Books, 1991.
Sacks, A. 'Sun, stealth and sand', *Evening Standard* 15 July 1998, 52.
Said, E. 'Reflections on Exile', in Ferguson, R., Gever, M.Minh-ha, Trinh T. and West, C. (eds) *Out There: Marginalization and Contemporary Culture* Cambridge, Mass.: MIT Press, 1990.
Savigliano, M.E. *Tango and the Political Economy of Passion* Boulder, Colorado: Westview Press, 1995.
Sawday, J. *The Body Emblazoned* London: Routledge, 1995.
Schick, I.C. *The Erotic Margin* London: Verso, 1999.
Snaith, Y. *Yolande Snaith Theatredance Blind Faith* video interview with Yolande Snaith. Spring re-Loaded 4 Video, The Video Place, London, 1998.

Soja, E. *Thirdspace* Cambridge, Mass.: Blackwell, 1996.
Spier, S. 'Engendering and composing movement: William Forsythe and the Ballett Frankfurt' *Journal of Architecture* Vol. 3 Summer 1998, 135–46.
Stallybrass, P. and White, A. *The Politics and Poetics of Transgression* New York: Cornell University Press, 1986.
Stallybrass, P. and White, A. 'From carnival to transgression [1986]', in Gelder, K. and Thornton, S. (eds) *The Subcultures Reader* London: Routledge, 1997.
Sulcas, R. 'William Forsythe. The Poetry of Disappearance and the Great Tradition' *Dance Theatre Journal* 9:1 Summer 1991a, 4–7, 32–3.
Sulcas, R. 'Frankfurt Ballet' *Dance and Dancers* August 1991b, 37–9.
Sulcas, R. 'Space and Energy' *Dance and Dancers* April 1992, 13–17.
Sulcas, R. 'Kinetic Isometries: William Forsythe on His "Continuous Rethinking of the Ways in which Movement Can be Engendered and Composed"' *Dance International* Summer 1995a, 8.
Sulcas, R. 'Summertime in France', *Dance Now*, Autumn 1995b, 37–41, 43.
Tarasti, E. 'On Post-Colonial Semiotics', in *European Journal of Arts Education* 2:2 1999, 69–83.
Tasker, Y. *Spectacular Bodies* London: Routledge, 1993.
Tharp, T. *Push Comes to Shove* NY: Bantam Books, 1992.
Thomas, H. (ed.) *Dance in the City* London: Macmillan (now Palgrave), 1997.
Urry, J. *The Tourist Gaze* London: Sage, 1990.
Van Kerkhoven, M. 'The Dance of Anne Teresa De Keersmaeker' *The Drama Review* 103 Fall 1984, 98–104.
Vickers, G. 'Manhattan Transfer' *World Architecture* Issue No. 10, 1991, 62–7.
Virilio, P. *The Lost Dimension* transl. D. Moshenberg New York: Semiotext(e), 1991.
Welton, D. (ed.) *The Body, Classic and Contemporary Readings* Oxford: Blackwell, 1999.
Wesemann, A., 'Body-Politics: Post modernism and Political Correctness' *Ballett International/Tanz Aktuel*, August/September, 1995, 44–7.
West, M. 'Reviews V. Martha Ullman West' *Dance Magazine* April, 1980, 44–6, 48–51.
Wheale, N. (ed.) *Postmodern Arts* London: Routledge, 1995.
Whitford, M. *Luce Irigaray Philosophy in the Feminine* London: Routledge, 1991a.
Whitford, M. *The Irigaray Reader* Oxford: Blackwell, 1991b.
Wilton, T. *Lesbian Studies: Setting an Agenda* London: Routledge, 1995.
Wolff, J. 'The culture of separate spheres: the role of culture in nineteenth-century public and private life', in Wolff, J. and Seed, J. (eds) *The Culture Capital: art, power and the nineteenth century middle class* Manchester: Manchester University Press, 1988.
Wolff, J. 'On the Road Again: Metaphors of Travel in Cultural Criticism' *Cultural Studies* Vol. 7 No. 2, 1993, 224–239.
Wright, E. *Feminism and Psychoanalysis. A Critical Dictionary*, Oxford: Blackwell, 1992.
Zizek, S. (ed.) *Mapping Ideology* London: Verso, 1994.
Zurbrugg, N. *The Parameters of Postmodernism* London: Routledge, 1993.

Index

49 Bis (Denizot 1992), 209
ability, 3, 6, 7, 166, 170–1, 181
abject, the, 17, 145, 149, 153, 155, 161, 171, 172, 173, 174
abjection, 140, 143, 150, 153, 154–5, 160, 161, 166, 171
Accumulation (Brown 1972), 158
Across Your Heart (Claid 1997), 23, 162, 165–6, 170–1, 173–4, 174–5, 177–8, 179
active/passive binary, 78, 92, 95
Aeroplane Man (Jonzi D 1997), 112–13, 114, 115–17, 118, 120, 121–2, 123, 124, 129–30, 131–2, 133, 134–5, 207
Aggiss, Liz, 23, 162, 164–5, 167–9, 174, 176–7, 179–80, 181, 207
Agis, Gabi, 48, 209, 210
Albright, Ann Cooper, 165
Anatomy Lesson of Dr. Tulp, The (1632), 139, 150, 151
Anderson, Lea, 22, 23, 34, 59–74, 139, 140, 144–5, 207, 211
androgyny, 69, 82, 87, 93, 168, 169, 198
animal/human boundaries, 165
animated space, 81, 96
  between bodies of dancers, 94
animating the space between, 91
animation of the body/space interface, 148
anti-Cartesian
  challenges to masculinized thought, 151
  perspective, 140
anti-Cartesianism, 142
anti-dualistic
  ideas, 140
  theories, 163
architecture, 12–13, 22, 31, 97, 100, 101, 105, 106, 109, 122–3, 184, 185, 188, 189, 191, 193–4, 195–6, 202–3, 205, 206

of choreography, 204
of disappearance, 201, 204
see also Libeskind, Daniel
Vitruvian, 185
assemblage, 79, 86, 88, 90
assemblages, 18, 78, 84, 91, 96

Baby Baby Baby (Anderson 1986), 71
Baggett, Helen, 165, 173, 180
Bakhtin, Mikhail, 17–18, 23, 71, 131, 162–4, 165, 166–7, 168–9, 170, 171, 172–6, 177–9, 180–2, 212
ballet, 10, 180, 202
  aesthetic, 185
  Romantic, 38
  style, 200
  vocabulary, 185, 187, 199, 204
Ballet Frankfurt, 23, 186, 194, 202
balletic bodies, 187
Banes, Sally, 45–6, 156, 178, 210
Baroque, the, 141, 142, 149, 155, 156, 158, 161
  folds, 18, 147, 151, 152
  paintings, 23, 146, 159
Barthes, Roland, 159
Baryshnikov, Mikhail, 144
Battersby, Christine, 161
Baudoin, Patricia, 190, 199, 200
Bauhaus, the, 193
Bausch, Pina, 8, 9, 12, 16, 18, 209
Beach Birds (Cunningham 1992), 68
beaches, 4, 59, 60, 61, 63, 64
becoming, 78, 80, 85, 88, 89, 91, 175
  animal, 86, 91, 95
  bodies, 18, 77
  body of the people, 181
  identity as a process of, 114, 120–1
  imperceptible, 87, 95
becomings, 79, 86, 87, 91, 202
being, 78, 114, 120–1, 184
Bentham, Jeremy (1748–1832), 125
Berger, John, 39

Berghuijs, Martine, 81, 208
Between/Outside (Power 1999), 78, 81, 83, 85, 88–9, 90–1, 92, 93, 94, 95, 207
Bhabha, Homi, 15, 17–18, 98, 99, 100–1, 102, 104, 107, 108, 109–10, 127, 132
Bharata Natyam, 97, 98, 99, 102–3, 104, 105, 107, 110, 111
binaries/binary oppositions, 9, 10, 14, 18, 20, 35, 78, 80, 89, 109, 140, 142, 163, 173, 181, 212
   blurred, 69, 73, 74, 78, 91, 92, 94, 95, 97, 100, 110, 167, 169, 171
   body/mind, 140, 141, 156, 180, 203
   challenges to, 166, 168, 170, 173, 205
   colonizer/colonized, 100
   East/West, 97, 100
   presence/absence, active/passive, 92
   private/public, 44
   subject/object, 66
binary
   coding, 143, 173
   extremism, 164
   see also dualisms
   thinking, 17, 18
Black Atlantic, 15, 112, 135
Blake, Steve, 35, 61
Blind Faith (Snaith 1998), 23, 139, 140, 142, 143, 144, 145–6, 148, 150–2, 153–4, 155, 156, 157, 159–61, 207, 211
Blueprint (Monk 1967), 45–6, 210
bodies
   and architecture, 105, 109
   and spaces, mutual construction of, 51, 57, 60, 106–8
   as borderline concepts, 15
   as source of folds, 141, 155
   becoming, 77, 181
   blurred, 15, 203
   boundaries of, 87, 139, 154, 173, 182
   bounded, 9–11
   Cartesian, 148, 151
   classical, 165, 175, 199
   connections between, 18, 79, 89, 90–1, 95
   contained, 82, 161
   docile, 129, 121, 125, 130
   double, 17–18, 167–9
   eroticized, 63–5, 73
   fragmented, 201, 204
   grotesque, 163, 165, 170–2, 174–6, 178–9, 181–2
   hybrid, 77
   in-between, 14–16, 20
   inscription of, 127
   lesbian, 77
   limits of, 3, 6, 60
   materiality of, 82, 139, 145, 146–7, 150, 151–2, 157, 161, 162, 163, 180, 181
   migrant, 77
   permeability of, 145, 153, 154, 166, 167, 171, 182
   transgressive, 162
   without boundaries, 92
bodies and souls, 141, 143, 149, 150, 161
bodily
   experience, direct, 193, 203, 205
   fluids, 17, 153, 156, 161, 172
   orifices, 3, 178
   space, 211
Body of the Dead Christ in the Tomb, The (1521), 154
body/soul
   connectivity, 140, 142
   dualism, 149
body/space interfaces, 1, 2–4, 13, 19–20, 77, 80, 148, 161, 183, 205
borders, 17, 100, 104, 113, 114, 132–4, 154, 158
   as social constructions, 67, 113, 133
   coastal, 22, 60, 66, 73
   fluid, 45, 67–8, 101
   of bodies, 21, 148, 153, 161
   of bodies and space, 3–4, 23, 43, 77, 139
   see also boundaries
Bordo, Susan, 9, 142–3, 151, 153, 157, 161, 184, 189–90, 192, 195, 200–1

boundaries, 15, 16, 36, 41, 65–8, 133, 153, 158, 196, 205
  between performers and spectators, 131
  blurred, 3, 10, 15, 17, 19, 20, 22, 44, 45, 46, 68, 69, 72, 74, 80, 100, 104, 110, 126, 134, 151, 169, 189, 205: of bodies, 2, 10, 17, 21, 41, 51, 89, 91, 173; of bodies and architecture, 195, 206; of borders, 114; of East West divide, 99; of insides and outsides of bodies, 147–8, 149, 184, 195; of inside/outside spaces, 161, 195; of private/public spaces, 94–5; of space and subjects, 200; of subject and object, 154; *see also* merging
  challenges to, 166–71
  coastal, 60
  fluid, 3, 16, 22
  of bodies, 3, 17, 20, 21, 41, 51, 79, 92, 143, 175, 182
  see also borders
  spatial, 4, 19, 20, 57
Braidotti, Rosi, 16, 98–9, 103, 106, 109, 110–11
Braun, Terry, 43, 99, 207, 211
Broken Column, The (1944), 170
Brossard, Nicole, 87
Brown, Trisha, 23, 44, 48, 56, 57, 139, 140, 144, 146–8, 152, 155–8, 160
Burgin, Victor, 159
Butler, Judith, 80–1, 89, 124, 125, 177, 197, 210

CandoCo, 2, 23, 165
Car (Anderson 1996), 61
Carnets de Traversée Quais Quest (Charlebois/Vasselin 1989), 32–3, 35, 39, 40, 41, 42, 209
carnival, 168, 170, 176, 181, 212
  excess, 172
  medieval, 17, 165
carnivalesque, 71–2, 162–4, 165, 166, 167, 169–70, 172–3, 176–7
  ambivalence of, 166, 168

  excess, 182
  parody, 175–8, 177
  performance, 178
Cartesianism, 184
Cartesian
  body, 150, 151
  dualism, 155, 193
  epistemology, 192
  masculinization of thought, 152, 158
  philosophy, 140, 141, 153, 157
  subject, 154, 161
  theory, 10, 143
chaos theory, 186, 191
Charlebois, Johanne, 32, 33, 207
Childs, Lucinda, 44–5, 47, 210
Cholmondeleys, The, 61, 64
  see also Anderson, Lea
chora, 154, 210, 211
Circumnavigation (Corsino, N and N 1992), 209–10
city spaces, 43–58, 105–11
Claid, Emilyn, 23, 81, 84, 92–4, 162, 165, 178, 208
classical architecture, 185
classical ballet, 179, 181, 185, 200
Claydon, Paul, 145, 146, 211
colonialism, 37, 41, 127, 130
  colonial power, 36
  colonial oppression, 113
  colonization, 37, 101, 209
  colonized and colonizers, 100, 132
connectedness, 111, 152
connectivity, 95, 141, 143
  of body and soul, 151, 159
  of corporeality and spirituality, 161
  of inside/outside spaces, 149
  within and between folds, 156
  with the world, 166
Connor, Steven, 29, 36
constructions
  mutual, of bodies and spaces, 4, 20, 21, 22, 45, 51–4, 56, 57, 59, 60, 64
  of identity, 114
  of nature, 41
  see also space, constructions of
contact improvisation, 50, 140, 156
Contact Quarterly, 147

containment, 123, 125, 130, 133, 148, 153, 156, 174
  see also bodies, spaces, subjects
corporeality, 53, 139, 143, 145, 146, 150, 152, 154–5, 157, 159, 160, 171, 194
Corsino, Nicole and Norbert, 209
Cowie, Billy, 164, 168, 207
Cross Channel (Anderson/Williams 1991), 32, 34–5, 38, 40, 41, 42, 61, 62, 207
Cunningham, Merce, 44, 68
cyborgs, 16, 60, 64, 66, 67, 69–72, 73, 74

Dancing in the Streets of London and Paris Continued in Stockholm and Sometimes Madrid (Tharp 1969), 47
Dandeker, Celeste, 165, 173, 174, 179
Dante, Alighieri, 201
Dark Hours and Finer Moments (Agis 1994), 210
Dartington College of Arts, 49
De Keersmaeker, Anne Teresa, 23, 49, 176, 183, 184, 188–9, 191–2, 193, 195, 197, 198, 200, 204–5, 208, 209
De Mey, Thierry, 188, 208
deconstruction, 121, 123, 126, 128, 131, 176, 183–4, 185, 186, 190, 194–5, 199–200, 204, 205
Deleuze, Gilles, 5, 18, 51, 77, 85, 87, 89, 91, 140, 141–2, 143, 147, 149–50, 151, 152–3, 154, 155–61, 186, 202, 210, 211
Deleuze, Gilles and Guattari, Felix, 18, 78, 79, 84, 92, 96
Denizot, Sarah, 209
Derbyshire, Charlotte, 165, 174
Deroute, La (Tafel/Jean 1990), 32, 33, 35, 39, 40, 41, 42, 57, 65, 207, 209
Derrida, Jacques, 198–9, 201, 203
Descartes, René, 9, 31, 140, 142, 192
  see also Cartesianism
desire, 10, 13–14, 18, 19, 22, 33, 77–96
  Freudian notion of, 78

lesbian, 22, 77–80, 89, 95
diaspora, 102, 108, 112
  diasporic artists, 98, 104
  diasporic immigrants, 119
  diasporic movements of peoples, 112
  diasporic spaces, 15, 22, 112
difference, 6, 7, 15, 22–3, 78, 80, 92, 98, 101–2, 109, 113, 114, 115, 117–18, 123, 124, 126, 129, 132, 134, 150, 151, 156–7, 170, 195, 209, 210
  bodies as markers of, 177
  celebration of, 23, 117, 134
  construction of, 113
differentiation, 134
disability, 23, 165, 170–1, 174–5, 178
disciplinary technologies of power – *see* technologies of power, 113
discontinuity, 186, 187, 189, 190, 199, 202–3, 204, 205
disembodiment, 181
disorientation, 190, 191, 194–5, 205
displacement, 16, 22, 29, 112, 113, 114, 120, 122, 132, 133, 135
distance, 8, 9, 13, 22, 31, 40, 41, 45, 84, 152, 157, 187, 191, 192, 194, 195, 204
Dogtown (Morris 1983), 23, 162, 165, 166, 169, 171, 172–3, 174, 176, 179, 181, 182
Dreyer, Carl Theodor, 139, 145, 149, 154, 159, 160
dualism, 31, 66, 140, 141, 142, 143
dualistic thinking, 9, 10, 19
Duets with Automobiles (Jeyasingh/Braun 1993), 22, 45, 97–111, 207, 209–10, 211
Dunn, Judith, 44
DV, 8, 210

edges, 62, 65–6, 200
  of bodies, 3, 23, 60
  of space, 3, 23, 60, 69
  of women, 80
Ellis Island (Monk/Rosen 1981), 46, 112, 113, 114, 115, 117–19, 120–1, 122–31, 132–5

Ellis Island immigration centre, 113, 115, 119, 130
  immigrants, 120, 132
  project, 121
  spaces of, 122, 125
embodiment, 8, 31, 40, 46, 72, 87, 161, 209
  embodied subjectivity, 42, 58, 73, 141, 143, 144, 148, 158, 163, 204, 205
Enemy in the Figure (Forsythe 1989), 23, 183, 186, 189–90, 192, 193, 196, 197, 203, 205
Enter Achilles (Newson 1995), 210
erotic, the, 22, 63, 70, 82, 84
eroticization, 59, 63–5, 68, 71, 73, 170
Euro-crash, 33, 49, 209
examination, 115, 127–30, 132
examination, medical, 124, 128
excess, 134, 166, 181, 185, 204
  bodily, 20
  of folds, 141, 142, 143, 158
  spatial, 20
excessive, 142, 161
  bodies, 17, 23, 148, 162
  flows, 182
  movements, 165
  overflows, 143, 171–5, 182
excessiveness, 171

Bausch, Deborah, 193, 194, 203
Featherstonehaughs, The, 61
  see also Anderson, Lea
Fehlandt, Tina, 176
feminine, 36, 54, 57, 65, 85, 167
  codes, 93, 197
  excess, 211
  gestures, 192, 197
feminine, the, 37, 41, 140, 141, 142, 143, 152, 153, 154–5
femininity, 37, 65, 197, 211
  construction of, 36, 52, 65
feminism, 60, 73, 196
feminist, 77, 113, 140, 161, 166
  architecture, 193
  project, 12, 104
  theorists, 16, 29, 30, 36, 40, 45, 48, 50, 65, 79, 98, 189, 203

theory, 7, 39, 60, 69
feminization, 64, 65
  of space, 210
fixed identities, 96, 134, 153, 171
fixity, 119, 163, 173, 187, 191, 195, 199
flaneur, 50, 57
flesh, 23, 63, 64, 80, 84, 85, 139–52, 154, 157, 159, 160, 161, 163
Florence, Penny, 5, 79
fluidity, 16, 22, 29, 40, 41, 43, 45, 46, 59, 60, 62, 67, 68, 70, 73, 78, 80, 88, 89, 94, 95, 96, 98, 101, 121, 142, 148, 150, 151, 152, 153, 156, 161, 163, 171, 190, 199, 205, 209
Fold, The (Deleuze, 1993), 141, 142, 150, 156
folded body, 150, 152
foldings, 18, 20, 140, 143, 156, 159
folds, 18, 23, 140, 141, 142, 143, 147, 149, 151, 152, 153, 155–60, 161, 182
Forsythe, William, 23, 183–205
Foucault, Michel, 4, 5, 9, 20, 51, 82, 113, 114, 121, 123–30, 151, 202, 204
fractals, 186, 191
Freefall (Agis/Bentley 1988), 48, 209, 210
French, Jon, 165, 174–5, 177–8, 179
Freud, Sigmund, 195
Friedrich, Casper David, 38
Fulkerson, Mary, 49

gaze, the, 12, 31, 39–40, 65, 87
  geographer's, 30–1, 35, 38–9, 41, 65
  male/masculine, 12, 13–14, 31, 39, 41, 42, 65
  power of, 38–40
  spectator's, 14, 66
  tourist, 39
gender, 21–2, 27, 28, 29–31, 35–40, 43, 48, 52–3, 54, 64, 69–71, 74, 77, 80–1, 142, 169–70, 173, 192, 197–8, 201
geometry, 185, 191, 201, 204, 205
  Euclidean, 189, 191, 199

geometric
  formations, 188
  grids, 185, 204
  seeing, 157, 192
  structure, 192
Gert, Valeska, 164, 176
Gilpin, Heidi, 190, 199, 200, 201, 202, 204
Gilroy, Paul, 15, 16, 112, 114, 118, 122, 131
Gottschild, Brenda Dixon, 116
Gough, Orlando, 99, 107
Goulthorpe, Mark, 186, 203
Gross, Elizabeth, 149, 150, 153
  see also Grosz, Elizabeth
Grosz, Elizabeth, 5, 10, 11, 14, 15, 16, 18, 19, 20, 46, 51, 52, 53, 57, 77, 78, 79, 82, 83, 84, 85, 86, 89, 90, 91, 92, 95, 119, 153, 154, 155, 160, 211
grotesque, the, 17, 18, 23, 162, 164–5, 167, 170–1, 172, 173, 175, 177, 178
  body, 162–3, 165–6, 167, 170, 174–5, 176, 178, 181–2
  movements, 166
  realism, 17, 163, 172, 179
  style, 168, 179
Grotesque Dancer (Aggiss 1987 revived 1998), 23, 162, 164–5, 166, 167, 168, 169, 171, 174, 176, 177, 179, 207
Guattari, Felix, 77, 86, 89, 91, 155, 186

Haraway, Donna, 16, 40–1, 42, 60, 64, 66, 67, 69–73, 74
Harris, Rennie, 116–17
Holbein, Hans, 154
Holliday, Billie, 16
home, 50, 52, 58, 98–9, 103, 104, 105–6, 107, 108, 110, 111, 119
homely places, 105
Homeward Bound (Spanton 1997), 78, 81–2, 84–5, 86–7, 95, 207
hooks, bell, 113
horizon, 28, 38, 66
House, The, 115
Huilmand, Roxanne, 43, 49, 51–4, 57, 207
hybrid, 60, 62, 64, 102, 139, 149
  existence, 100
  identities, 110
  see also bodies, hybrid; spaces, hybrid
hybridity, 15, 17–18, 20, 22, 69, 98, 99, 101, 102, 106, 107, 109, 111
  see also spaces, of hybridity
hybridization, 16

identity, 3, 7, 15, 16, 22–3, 29, 87, 101, 104, 106, 112, 113, 114, 124, 134, 149, 157, 161
  ambiguisation of, 160
  as process, 120–2
  construction of, 117–23
  enunciations of, 120
  fixed, 10
  fluid, 111, 122, 141, 150, 153
  formation, 112, 113
  fragmentary, 118–9, 122
  lesbian, 77
  transitory, 88
If You Couldn't See Me (Brown 1994), 23, 139, 140, 143, 144, 146, 148, 149, 150, 152, 153, 156–8, 161
imaginary homeland, 98, 105–6
Imaginary Homelands (Rushdie 1991), 98
imagined spaces, 107
improvization, 116, 199, 202
in-between bodily fluids, 153
in-between spaces, 4, 14–17, 18, 22, 46, 60, 77, 87, 94–5, 97, 101, 102, 110, 111, 112, 113, 134, 161, 184, 202, 205
in-between, the, 154
in-betweenness, 113
inscription, bodily, 20, 51, 54, 57, 64, 82, 98, 107, 127, 130
inscriptive processes, 121
inside/outside, 3, 9, 10, 17–20, 36, 44, 45–6, 50, 54, 89, 91, 92, 107, 110, 133, 134, 139, 143, 147–8, 149, 153, 154, 158, 161, 167, 173, 181, 189
  bodies, 21, 23, 205
  bodily borders, 183

inside/outside – *continued*
  body boundaries, 162
  body interfaces, 165
  body/space borderlines, 182
  borders, 196
  boundaries, 182
  merging, 95
  spaces, 21, 23, 49
intensities, 18, 79, 80, 81, 84, 90, 91, 96, 160
interconnectivity, 16, 18, 21, 22, 72, 78, 84, 85, 89, 99, 100, 104, 110, 111, 172, 178, 193, 205, 206
interfaces, 19, 46, 81, 95, 104
  between bodies, 18, 77, 83
  between bodies and cities, 58
  between bodies and coastal environments, 73
  body/space, 78, 80
  inside/outside, 17–20, 165
Irigaray, Luce, 19, 39, 77, 78, 79, 80, 84, 85, 87, 89, 92, 93, 95, 196, 197

Jam (Tharp 1967), 48
Jameson, Frederic, 5, 29
Jean, Roderigue, 33, 207
Jewish Museum, Berlin, 190, 196, 202–3
  see also Libeskind, Daniel
Jeyasingh, Shobana, 22, 97–111, 207, 209–10, 211
Joan (Anderson/Williams 1994), 23, 139, 140, 142, 143, 144–5, 148–9, 152, 153, 154, 155, 156, 158, 159, 160, 161, 207
Joan of Arc, 23, 150
Johnson, Kwesi, 165, 174, 179
Jones, Bill T., 144, 157, 211
Jones, Stuart, 165
Jonzi D, 22, 112, 114–7, 118, 120, 121–2, 129, 131–2, 133, 135, 207
jouissance, 154, 211
Joyce, Heather, 82, 92, 93, 94
Judson Dance Theater, 44, 50, 147
Judson Memorial Church, 44
Judson Gallery, 45
Just Before (De Keersmaeker 1997), 192

Kael, Pauline, 145
Kahlo, Frida, 170–1, 174
Kaprow, Allan, 44
Kent, Sarah, 36, 37
Kolb, Wolfgang, 43, 49, 53, 207
Kristeva, Julia, 17, 140–1, 143, 145, 153, 154–5, 160, 161, 166, 171–3, 196, 211

Lacan, Jacques, 11, 79, 195, 196
Lament of the Empress (Bausch 1989), 8, 9–10, 11–12, 13, 15, 16, 17, 18, 19, 39, 45, 209, 210
Land Jäger (Schneider 1990), 32, 34, 35, 37, 41, 42, 209
landscape, 33, 35, 37, 42, 62, 64–5
Last Supper, The (1498), 139, 145, 146, 153, 160
Lefebvre, Henri, 4, 5, 11–13, 20, 30–1, 43, 56, 59, 63, 71, 81, 200
Leibniz, Gottfried, Wilhelm, 141–2, 147, 155
Leonardo da Vinci, 139, 146
lesbian performance, 93
  see also bodies, desire
Libeskind, Daniel, 185, 190, 193, 196, 200, 202–3, 204
  see also Jewish Museum, Berlin
libidinal space, 23
Limb's Theorem (Forsythe,1990), 186, 187
linearity
  explosion of, 190, 195, 205
  of the classical body, 199
  of logical structures, 200
  of perspective, 191, 196
Lingis, Alphonso, 129
linkages
  between bodies, 79, 89, 91
  in relation to lesbian desire, 95
Livet, Anne, 45
locatedness in space/time, 192, 195
London Contemporary Dance School, 115
Longo, Jovair, 145
Lorraine, Tamsin, 79, 87, 96
Lovell, Thomas, 186
Lyotard, Jean-François, 209

Lyrikal Fearta, 115
  see also Jonzi D

machinic
  alliances, 79, 90
  assemblages, 88–92, 95
  connections, 80, 85–6
  enunciations, 92
Madden, Drostan, 148
Making of Maps (Jeyasingh 1992/3), 101, 108, 110
Man Walking Down the Side of the Building (Brown 1969), 48
Manet, Edouard, 79
maps and mapping, 5, 27, 29, 36–7, 40, 65, 68, 101, 152, 191, 205, 209, 211
marginalization, 105, 164
marginality, 113, 163
margins, 36, 113
Mark Morris Dance Company, 165
Marks, Victoria, 2, 207
masculine, the, 140, 143, 155, 167
  Cartesian subject, 156, 161
  codes, 93
  gaze, 65, 94
  knowledge, 31
  mode of thinking, 36
  perspective, 39, 41, 42, 65, 189
  public spaces, 97
  qualities, 152
  viewpoint, 38, 184
masculinity, 31, 37, 203, 209
masculinization of thought, 142, 151, 152, 157
Massey, Doreen, 5, 29, 30, 35, 36
materiality *see* bodies, corporeality, flesh
McDonagh, Don, 210
Medley (Tharp 1969), 48
merging
  of bodies and space, 3
  of/with bodies and architecture, 191, 192
  of self and world, 169
  of spaces and subjects, 204
  see also boundaries, blurred
Merleau-Ponty, Maurice, 19
migration, 30, 32

migrants, 38, 42
  see also bodies, migrant
Miller, Graeme, 145
mimicry, 128, 131–2, 170, 171, 175, 176
mind *see* binary body/mind
mirror stage, 195
Mlle Victorine en costume d'Espada, (1862), 79
Monnaie, La, 188
Monk, Meredith, 22, 44, 45–7, 50, 112, 114–5, 123, 128, 129
Monk by the Sea, The (1808–10), 38
Morphologies (1996–98), 210
Morris, Mark, 23, 162, 165, 169–70, 172, 179, 182
Morris, William, 193
multicentricity, 199
multiplicity, 18, 56, 70, 79–80, 88, 91, 92, 141–2, 156, 158, 163, 189
Mulvey, Laura, 5, 12, 39, 65
Muurwerk (Huilmand/Kolb 1987), 43, 45, 46, 47, 48, 49, 50, 51–4, 57, 106, 207

nationality, 119
nature, 37, 38, 40–1, 42, 64–5, 155
negative space, 202
Newson, Lloyd, 210
Nietzsche, Friedrich, 51
Nodine, Rick, 145
nomadism, 27, 99, 104, 110
  see also subjectivity, nomadic

object positions, 93
objectification, 39, 40, 41, 66, 94, 123, 126, 128, 144, 149, 157–8
objects
  beings as, 198
  body as object of science, 150–1
  constructed as, 118
  of desire, 93
  of the masculine gaze, 39–40, 42
  viewed as, 45
ocularcentrism, 198, 204
off-balance, 187, 194, 199
Oldenburg, Claes, 44

One Two Three (Tharp 1967), 48
Ono, Yoko, 165, 169, 172, 176
Opal Loop (Brown 1980), 158
openness, 200
  of the double body, 168
  of the grotesque body, 162, 166, 171
  of the subject, 173–4
origins of identity, 114, 117–19
orthogonal space, 190
other, the, 28, 37, 38, 41, 78, 89, 117, 157, 192
othering, 117–18, 132
otherness, 125, 141, 157, 195, 202
  constructions of, 42
others, 54, 73, 113, 118, 120, 132–4
Out on the Windy Beach (Anderson 1998), 22, 59–74, 168, 210
Outside/In (Marks/Williams 1995), 2–4, 17, 207

Palermo Palermo (Bausch 1990), 209
panopticism, 125, 128
panopticon, 125, 203, 204
Panzen, Jerry, 115, 123
Paradise (1588), 159
Parker, Alice, 87
Parody, 63, 93, 163, 165, 166, 167, 175–8, 181, 197, 204, 209
PARTS (Performing Arts Research and Training Studio), 188
Paxton, Steve, 49–50, 146–7, 148, 149, 152, 156, 157, 160
perception, 9, 45, 52, 56, 140, 141, 148, 187, 189, 193–4, 201, 205
  corporeal, 157
  embodied, 46, 48
  of the audience, 47, 48
Perfect Moment (Anderson/Williams 1992), 62
performativity, 15, 80–1, 89, 197, 210
periphery, 36
permeability *see* body, permeability of
Perron, Wendy, 156
perspectival spaces, 191
perspective, 8, 10, 11–13, 14, 19, 31, 35, 48, 107, 157, 184, 185–6, 187, 189–190, 196, 198, 199, 200–1, 203

phallocentrism, 198, 204
phallus, 196–7
Picasso, Pablo, 12
place, 15, 28, 31, 35, 38, 42, 50, 51, 54, 55, 56, 57, 108, 113, 114, 117–18, 120, 122–3, 211
Plato, 210, 211
plural sexuality, 79
polycentricity, 189–90, 199, 201, 203–4
polymorphous
  figures, 199
  perversity, 80, 86, 92–4
  subjectivity, 74
positive space, 202
post-colonial
  artist, 132
  criticism, 29
  subject, 117
  theories, 7, 16, 17, 98
  theorists, 15, 112, 127, 133
postmodern
  theory, 5, 6, 7, 27, 29
postmodernism, 36, 163, 212
postmodernity, 71, 209
poststructuralism, 14, 29, 184, 186, 198
post-structuralist
  theory, 5, 7, 8, 20
poststructuralists, 77, 140, 151
Powell, Sandy, 62, 69
Power, Lucille, 81, 83, 88–90, 93, 94, 207
presence/absence binary, 10, 78, 92, 95
proprioception, 194–5, 199, 201, 204

Quinn, Marc, 210

Rabelais, 162–3, 166–7, 169, 174, 177
Rabelais and his World, 162
rap music, 22, 112, 114, 116, 118, 120, 121, 131, 132, 133
Rauschenberg, Robert, 144, 210
Recent Ruins (Monk 1979), 115, 126
rectilinear, 201
rectitude, 201
re-embodiment, 181, 182

reinscription, 53–4, 56, 57, 106, 109
Rembrandt, 139, 150, 151
  see also 'The Anatomy Lesson of Dr. Tulp'
Renaissance, The, 22, 23, 139, 143, 146, 154, 159, 163, 180, 189, 198
representation, 6, 7, 87, 102, 184
Requel, The, (Jonzi, D 1997), 115
Reservaat, (Van Gool 1988), 78, 81, 82, 83, 85–6, 88, 91, 94–5, 207, 210
resistance, 111, 113–14, 124, 126, 129, 131, 134
Resto, Guillermo, 176
rhizomatic experimentation, 186, 191
Roof Piece (Brown 1971), 48
Rosas Dance Company, 49, 188, 208
Rosas Danst Rosas (De Keersmaeker 1983 film version 1997), 23, 45, 183, 184, 191, 193, 194, 195, 197, 203, 205, 208
Rose, Gillian, 5, 27, 30–1, 35, 38, 65
Rosen, Bob, 115, 211
Roy, Sanjoy, 102
Rushdie, Salman, 98

Said, Edward, 133
sameness, 117, 150, 151, 156–7, 170
Savigliano, Marta, 82
Schneider, Stefan, 34, 207
scopic regime, 198
self/other binary, 10–11, 13, 14, 22, 31, 78, 94, 95, 96, 117, 142, 143, 150, 154, 167, 169, 173
self-subjectification, 125
separation from the world, 157, 184, 192–5, 199
sexuality, 3, 7, 19, 23, 59, 77, 81, 93, 166, 169–71, 181, 209, 210
Singh-Barmi, Kuldip, 165, 166, 173, 174–5, 179
Smit, Pépé, 81, 208
Smith, Sue, 165
Smithereens, The (Anderson 1999), 62
Snaith, Yolande, 23, 43, 49, 57, 139, 140, 145–6, 150, 157, 160, 207, 211
Snapshots from the City (Oldenburg, 1959), 44

Soja, Edward, 125, 132
Solo (Forsythe 1985), 186, 194–5, 199–200, 203
space, 1–6, 8–11, 12–13, 28, 30–1, 35–8, 45, 56, 101, 113, 125, 183, 184, 194, 205, 210
  ambivalent, 15, 21, 163
  architectural, 23, 185, 204
  as a bounded entity, 210
  as constructed/construction of, 4, 12, 13, 16, 27, 35, 47, 50–1, 53, 59, 64, 67, 68, 78, 107
  decentered and fragmented, 191
  heterogeneous, 190
  homogeneous, 190
  liminal, 60, 113, 134
  limits of, 3, 60, 66, 67
  nostalgic, 36–7
  of bodies, 81
  organization of, 184
  private/public, 94–5, 101
  private, 10, 15, 19, 46, 50, 52
  public, 10, 15, 19, 31, 46, 48, 50, 52, 61, 88, 95
  static, 202
  transparent, 30, 31, 35–6, 39, 41, 42
  urban, 55–6
spaces, 122
  animated, 94
  between borders and boundaries, 104
  between dancers, 3–4, 7, 82, 88–89, 95, 110
  between dancers and architecture, 109
  between parts of bodies, 82, 84
  blurred, 64
  coastal, 59–76
  hybrid, 14, 98, 101
  interior and exterior, 94
  of ambiguity, 14, 17, 20
  of becoming, 14, 202
  of consumption, 71
  outdoor, 6–7, 27, 34, 47, 59
  positive, 202
  see also city spaces; urban spaces
Spanton, Sarah, 81, 82, 83, 84, 86, 87, 88, 90, 93, 207

spatial
  composition, 188
  containment, 114, 132–4
  division, 133
  experience, 189, 195
  interfaces, 77
  linkages, 91–22
  operation, 134
  organization, 126
  texts, 184
  texts of dance, 183
  treatment, 113
spatialities of subjectivity, 206
spatiality, 90–1
spatialization
  of desire, 77–96
  of subjectivity, 206
Spier, Steven, 189
Split Solo (Brown 1974), 158
Step in Time Girls (Snaith/Braun 1988), 43, 45, 46, 47, 48, 49–50, 54–6, 57, 106
Steptext (Forsythe 1984), 200
Street Dance (Childs 1964), 45, 47, 210
Street, The (Oldenburg 1959), 44
Stride (Tharp 1965), 48
subject, the, 154, 184, 192, 196, 198
  ambiguity of, 17, 55, 173–4
  ambivalent, 182
  in process, 171, 174, 204
  open, 174
  positions, 6, 93
  rational unifed, 9, 142, 143, 167, 171, 184, 185, 186, 192
  subject/object binary, 78, 87, 92, 95, 96, 143, 149
  visual and spatial construction of, 196
subjected bodies, 129
subjectification, 121, 124–6, 130, 131
subjection, 6, 124, 128
subjectivity, 6–7, 11, 20, 27, 42, 117, 140, 205, 206
  ambiguity of, 154, 163, 173–4
  constructions of, 4, 6–7, 9, 11, 19–20, 22, 23, 41, 43, 45, 56, 59, 64, 77, 124, 143, 184, 195–8, 206

  embodied, 20, 41, 42, 58, 74, 142, 143, 158, 161, 204
  female, 104, 106, 111, 140, 143
  female feminist, 99, 103
  feminine, 57, 158, 161
  fluid, 6, 41, 62, 111, 134, 163
  multiple, 163, 169, 209
  nomadic, 16, 98, 99, 102–4, 106, 108–9, 110, 111
  non-fixity and instability of, 18
  subjects, 118, 123, 125
  contained, 113, 171, 204
Sulcas, Rosalyn, 187, 192, 193, 198
surface(s), 80, 83, 100, 105, 156, 160, 181
surveillance, 118, 125, 127–8, 129
  camera, 125, 134

tableaux, 115, 119, 129, 146, 151, 153, 157, 159, 161, 181, 211
Tafel, Tedi, 33, 207
technologies of power, 9, 114, 121, 123–30, 135
territorialization, 114, 132–4
Tharp, Twyla, 44, 47, 48, 53
The Passion of Joan of Arc (1928), 139
Thelma and Louise (1991), 29
Tintoretto, 159
Topic II (Denizot 1992), 209–10
touch, 2, 19–20, 81, 83–5, 88, 90, 93, 95, 102–3, 105, 109, 110
touching, 77, 79, 80, 89, 92, 94, 96, 192, 196–7
transformation, 18, 72, 77, 79, 80, 85–7, 89, 91, 95, 130, 140, 146, 147, 150, 161
transgression, 165, 169–70, 179, 182
  of boundaries, 163, 168, 196
transmobility, 99, 104, 110
transparency, 37, 40, 65
Trigg, Russell, 145, 150, 211
two lips, 19, 78, 79–80, 85, 89, 95

urban, 35, 96
  environment, 45, 46, 52, 56
  landscape, 107
  sensibility, 44, 47
  spaces, 50–1, 55, 56

urban – *continued*
  surfaces, 48
urbanity, 23

van der Velde, Henry, 188, 191, 193, 194, 196
Van Gool, Clara, 81, 208, 210
Van Kerkhoven, Marianne, 191, 197
Vandekeybus, Wim, 209
Vasselin, Harold, 32, 207
Vesalius, Andreas, 146, 149, 150–1
Vessel (Monk 1971), 46–7
Virginia Minx at Play (Claid 1993), 78, 81–2, 84, 92, 95, 208
Virilio, Paul, 190
visual continuum, 189, 191, 195, 203
visual, dominance of the, 196–7, 198, 202
visual markers, 198

visualization, the logic of, 12, 14, 19, 23, 31, 144, 157–8, 182, 184–5, 191–2, 195, 198, 199, 200, 204, 205
visualization, 10, 12, 37, 189, 198, 200–4
visualization of space, 184

Walking on the Wall (Brown 1971), 48
Willems, Thom, 186, 187
Williams, Margaret, 2, 34, 144, 207
Wilson, Kepple and Betty's *Sand Dance*, 63, 71–2
Wimbledon School of Art, 211
Wolff, Janet, 5, 27, 29, 37, 40, 50

Yippeee!!! (Anderson 2006), 210
You Can See Us (Brown 1995), 158, 211